THE LANGUAGE OF LIFE

THE LANGUAGE OF LIFE

HOW COMMUNICATION DRIVES HUMAN EVOLUTION

JAMES LULL
AND EDUARDO NEIVA

59 John Glenn Drive
Amherst, New York 14228–2119

Published 2012 by Prometheus Books

Cover design by Jacqueline Nasso Cooke

Inquiries should be addressed to
Prometheus Books
59 John Glenn Drive
Amherst, New York 14228–2119
VOICE: 716–691–0133
FAX: 716–691–0137
WWW.PROMETHEUSBOOKS.COM

16 15 14 13 12 5 4 3 2 1

Library of Congress Cataloging-in-Publication Data

Lull, James.
The language of life : how communication drives human evolution / by James Lull and Eduardo Neiva.
p. cm.
Includes bibliographical references and index.
ISBN 978-1-61614-579-8 (pbk. : alk. paper)
ISBN 978-1-61614-580-4 (ebook)
1. Human evolution—Philosophy. 2. Social evolution—Philosophy. 3. Communication. 4. Language and languages. I. Neiva, Eduardo. II. Title.

GN281.4.L85 2012
599.93'8—dc23

2012000379

Printed in the United States of America on acid-free paper

For Bunchie and Dude

CONTENTS

INTRODUCTION

You are the leader of a primitive village. If you want to survive in a hostile world, you must evolve! In *The Adventures of Darwin*, you will drive the evolution of your village from a small group of simple primates to a powerful, intelligent colony of humans. Lead your tribe on adventures, teach them to hunt, teach them to build, and teach them the simple power of language. . . .

—description of *The Adventures of Darwin* on http://usplaystation.com

The Sony PlayStation® video game *The Adventures of Darwin* encourages gamers, many of whom are in their early formative years, to see the world through evolutionary eyes.[1] In a creative and cogent way, the game highlights emergent forms of ancient social organization, the empowering force of modern communication, and the ability of our species to self-consciously change conditions for the better. Over the millennia, our ancestors developed an unequaled ability to share information, form and maintain social relationships, innovate, and build advanced civilizations because they became the most skilled communicators on the earth.

From the moment the first single-cell microorganism split off from its bacterial host more than three billion years ago to the latest political and cultural upheavals like those taking place around the world today, life on the earth derives from ceaseless adaptation to constantly changing conditions. The natural world is perpetually transformed by means of an innate and complex interconnectedness that functions timelessly within and among all living forms. Evolution is the continual process of change and the diversification of all living things. Communication processes—the sending of signals and exchange of messages within and between organisms—shape the evolutionary adaptations that take place.

The speed and impact of the political and cultural uprisings in the Middle East and North Africa in 2011, for example, were made possible by the ability of communications media to instantly grow networks of people who share a common goal—the dismantling of repressive political regimes—and common value—the desire for greater freedom and social justice. The spread of information and exchange of messages reinforced and encouraged the dissenters and provoked an unprecedented crisis of governing authority in the region. The parade of political protest—in Iran, Tunisia, Egypt, Bahrain, Yemen, Libya, Syria—grew to resemble what biologist Stuart Kauffman calls "the adjacent possible"—when conditions are right, motivated contact between contiguous organisms or events can accumulate to bring about extraordinary change.[2]

As the rebellion in Egypt swelled, Wadah Khanfar, director-general of the Arabic-language satellite television channel Al Jazeera (he stepped down from that position in 2011), coined the term "informational ecosystem" to describe the resonance that developed between traditional and new media. The ground had already been prepared. Conditions conducive to the Arab Spring uprising—the name given to the wave of prodemocracy rebellions and protests—had been created by the way mass media, the Internet, and mobile communications technologies altered the cultural landscape over the past decade. Satellite television transmitted fresh ideas to the Middle East. New media technology was firmly in place. The messages that flowed from Twitter®, Facebook®, YouTube®, and other social networking sites then jump-started the revolt and created spontaneous communities of protesters, especially among disaffected members of the young urban middle class.

As the number of online protesters grew, more and more people took to the streets. But the Internet and personal communications technology alone could not facilitate a critical mass. Only about 20 percent of Egyptians have access to the Web, and less than half own a cell phone. More than a quarter of the population is illiterate. It was round-the-clock coverage by Al Jazeera, Abu Dhabi, CNN, BBC, and other regional and international satellite television channels that provided images and reports that directly or indirectly reached nearly everyone in the country and beyond. Egyptian state media began to get nervous as many reporters, anchors, and managers openly sided with the revolutionaries. Reporters and editors from all these channels stood up for freedom despite the harsh repression.[3]

The impact of satellite television, the Internet, social media, and mobile phones smashed any pretense of a state monopoly on information. Governments could only attempt to block websites, shut down phone service, sack television offices, confiscate equipment, ban media coverage, arrest journalists, and blame foreigners for stoking unrest. Popular resistance and collective action were being fueled by the power of shared information and mediated social interaction. The bottom-up, decentralized dynamic of change was in full effect.

Communications media and information technology didn't cause the abrupt insurrection in Tunisia, Egypt, Yemen, Libya, Syria, and other neighboring countries in 2011. The people in that region had long suffered corruption, repression, and abuse. But when poor Tunisian vegetable seller Mohamed Bouazizi set himself on fire in public to protest bribes routinely extracted by government officials, powerful images of his charred body and a compelling narrative that matched people's everyday experience started to circulate. A firestorm of political protest began to burn. Tunisian dictator Zine el-Abidine Ben Ali was chased out of the country. Unrest spread to Egypt when Google executive Wael Ghonim launched a Facebook page showing a grotesquely disfigured young Egyptian man beaten to death by police. Encouraged by the overthrow of Tunisia's president, weary of corrupt autocratic rule, and sparked by an intoxicating revolutionary spirit circulating through mass and social media, many Egyptians took decisive action.

Provocative images and videos were posted online. Satellite TV reporters swarmed into Cairo. Social media and mainstream media fed content to each other nonstop. Twitter's immediacy, reach, and ease of use allowed people to express their outrage and plan subversive activities. Facebook's distributive capacity and impromptu social networks delivered a double dose of soft-power influence. Trust between unacquainted social actors online and on the street grew stronger by the hour. Dictatorial power crumbled under pressure being applied by a social movement that formed spontaneously inside the flexible regimes of information and communication. The aggregate force of human determination and communications technologies pushed the power of autonomous connectivity past the tipping point of authoritarian control.

Transformative events like those in the Middle East and North Africa occur because individual people courageously take steps to inspire change. But their actions stand to benefit more than the individuals involved. People survive as social groups. Humans have obsessively developed faster and faster ways to create, connect, and work together for good reason. Human social interaction is uniquely collaborative and productive.[4] And, as science journalist Nicholas Wade points out, "Little is more critical to the survival and development of a social species than its members' ability to communicate with one another."[5] Entire civilizations arose because of the advantages brought about by increasingly advanced forms of cooperative communication that Wade says mark "perhaps the only clear distinction between people and other species."[6]

It wasn't always so. A fortuitous progression of biological mutations, physiological adaptations, and behavioral changes eventually endowed humans with the ability to express themselves in complicated ways and to coordinate their efforts to survive. Simple gestures and vocalizations were the first means through which humans' growing ability to imagine, create, and reason could positively influence their social organization, problem solving, and productivity. The consequences were decisive. Growing communication skill helped change our species from scavenger and the prey of larger and faster beasts to that of innovator and successful predator.

In the following pages, we argue that biological and cultural life is driven by a vast and dynamic succession of communicative interactions that are ruled by common principles—an endless chain of communication. A more familiar image was planted firmly in the public mind long ago: the iconic "tree of life" trope that Charles Darwin and others have used to describe genealogical relationships among plants and animals. The tree of life metaphor became a valuable template for organizing and displaying biological classifications and relationships. But the number of branches, stems, and twigs required to represent evolution's complexity have proliferated to such an extent that now the tree's graphical grid has become nearly undecipherable, even by many scientists.[7] Immense natural growth, hybridity, and evolutionary doubling back make the graphic look in some ways more like a bushy shrub or impenetrable thicket than a majestic tree.

Darwin's crucial formulation "common descent with modification" had the effect of directing much subsequent analysis to a search for the roots and branches of the evolutionary tree—its taxonomic paths and classifications. Our focus is different. We concentrate on the dynamic spaces where life-forms evolve—the links that make up the chain of communication—particularly as they influence the social and cultural life of modern human beings. The shift in emphasis away from elemental roots and descendants toward complex connections and processes expands the dual evolutionary structure to a mediating triangle—mutation, selection, and communication.

Ubiquitous connectivity—the totality of communication processes that creates and sustains biological equilibrium and social stability—reveals the essential unity of life on the earth. Some contemporary observers, however, interpret connectivity and communication more narrowly. For instance, connectivity can be thought of primarily in technical terms: the role of the Internet and personal communications technology is to facilitate remote forms of social interaction. Or, connectivity can be regarded as the means by which institutional power is created and sustained. Sociologist Manuel Castells, for example, speaks of "switching power . . . the capacity to connect two or more [information] networks in the process of making power for each one of them in their respective fields."[8] In Castells's view, global communication functions mainly as a system of interlocking corporate, financial, cultural-industrial, technological, and political entities that benefit mutually from the arrangement.

In many respects, the focus on technologically facilitated social interaction and concerns about the hegemonic power of the global communications network ring true. But connectivity and communication play even more important roles in evolutionary processes. The influence of evolutionary communication extends from processes of cell-to-cell interaction and the transmutation of simple creatures from asexual to sexual beings all the way to the formation of complex societies by multicellular organisms and the conduct of everyday affairs routinely undertaken by ordinary people in today's Communication Age.

Evolution is best understood as an endless becoming wherein information flows continuously inside organisms and from organisms to the environment and back. The ways information moves around define and order life. In biological terms, the key to the survival and reproduction of sentient beings is the ability of the organism to successfully organize and transmit signals internally; to send messages to other organisms; and, for social species, to coordinate activity.

In social and cultural terms, the key to human development is the ability to imagine, innovate, exchange ideas, and pass along ideas from one generation to the next. From the first printing press, telegraphy, telephony, and photography through the electronic media that emerged in the last century and on to the Internet, social media, and mobile smart devices of the current era, communications technology opens up discursive spaces that can be used to challenge tradition and authority. Unlike animals that possess more limited communication ability, human beings can advantageously expand on what they learn from other groups and from their past experiences. The advent of new forms of communication and the rate of cultural development have increased steadily together since complex communication and diverse kinds of social interaction began to blossom during the late Pleistocene era, from about 125,000 to 10,000 years ago.[9]

Evolutionary principles underlie the rationale that supports the freedom of information and communication. To survive, you have to cooperate. To cooperate, you have to communicate. The United Nations' Universal Declaration of Human Rights speaks of the right of every individual person—"every*one*"—to seek, receive, create, and express ideas through communication channels that range from simple movements of the body and utterances in local contexts to sophisticated engagement with high technology in global communications. The UN confirmed and extended this line of moral reasoning in 2011 when it declared access to the Internet to be a human right. The modern guarantees are twofold: people should not only have the right to freely express themselves on the Internet, and through all other communication channels, but also to be exposed to the thoughts of others.

The technological landscape today makes it possible for people to originate, produce, and distribute ideas much more easily and with far greater impact than ever before. Unfortunately, the same technologies that allow people to resist authority, improve their cultural conditions, and exercise their expressive potential can also be used by a society's ideological and cultural power brokers to reinforce the status quo or to change things for the worse. While the United Nations Human Rights Council calls on governments to protect citizens' rights to speak freely and have access to ideas expressed by others on the Web, for instance, many state governments block websites that offend political or religious sensibilities. Facebook, Twitter, YouTube, and Google® have all hit major roadblocks in China. Limited access and outright bans on politically sensitive or "culturally inappropriate" websites have been imposed. Pakistan disallows access to social network pages that post images of Muhammad. Under some of these governments, intellectual property rights and piracy laws are applied in ways that limit marketplace competition unfairly. The United States and other Western nations struggle to protect copyright holders without infringing on free speech. Authorities everywhere follow the electronic trails of activists to make arrests. The Web gives rise to other kinds of dangers. Religious fanatics and unscrupulous businesspeople are empowered. Hackers steal personal identities. Even some individuals who live comfortably in economically developed countries suffer debilitating psychological, social, and cultural effects from their immersion in modern communication technologies.[10]

The overarching truth, however, should not be obscured: throughout history communication technologies have had a positive impact overall, and the conditions for human development have never been more favorable than they are today. The ability to innovate culturally has allowed modern humans to adapt to virtually any circumstance, and now, more than ever, our skill as social communicators holds the key to our future success.

Despite the essential link that connects communication and human evolution, the massive body of empirical research that supports evolution, and the fact that it is one of the most powerful scientific explanations ever put forward, evolutionary theory has yet to impact the development of communication theory in any significant way. The true origins of human communication are left unexplored. Whether the field is microeconomics/macroeconomics, criminology, political science, or clinical psychology, social scientists rarely frame their discussions and debates within an evolutionary framework. They work within established theoretical categories that are very difficult to challenge.

This problem runs in both directions. Communication processes have never been adequately explained or properly integrated into the scientific narrative of human evolution. Thus we argue for a major infusion of Darwinian thinking into communication theory and for greater attention to the role of communication in evolutionary processes. Evolutionary communication is a powerful theoretical perspective that applies to all forms of biological and social interaction. It is not just a disciplinary subdivision existing alongside categories that sort the field according to the number of people involved in an interaction (interpersonal or group communication), to whether or not technology is present (mediated/mass communication), or if the interlocutors share or don't share a cultural background (intercultural communication).

Our goal as the authors of this book is to explain how communication drives biological and cultural change. To do so, we analyze core domains of human existence—sheer survival, sex, culture, morality, religion, and technological change—as communications phenomena. We hope the mosaic presented in the following pages will encourage scholars to investigate communication's role in evolutionary theory in ways that go far beyond what we do here. New modes of scholarly analysis are desperately needed. But we didn't write this book for academics. The book is a protest against conventional thinking, including the misplaced political correctness that has kept evolutionary theory on the margins in the social sciences. Our audience is the educated reader who is curious to learn more about human evolution and the communication processes that drive it. The book has personality; it is to be enjoyed. Darwin himself wrote books for the public, not just for the scholarly community. He filled his books to the brim with stories and personal asides. Darwin typically offered small theoretical explanations that he supported with a wealth of examples, insights, and critical commentaries that led eventually to dramatic conclusions. That's how evolutionary theory can be made clear and interesting for the reader—then and now. To make our work as accessible as possible to everyone, we include a glossary at the end of the book that succinctly defines the main concepts we take up in the chapters that follow.

Happily, this book fits within a broad pattern of today's Western media and culture industries, paying increased attention to the empirical reality of evolution and evolutionary theory. Television programs aired in recent years plainly explain human ancestry, the principles of natural and sexual selection, DNA and the human genome, the origin of tools and symbolic forms, and other critical evolutionary topics.

Visual and musical celebrations of evolution spring from Cirque du Soleil® and musician Björk. Baba Brinkman raps about evolutionary theory. Most zoos and museums tell the truth about how we got here. Bestselling books educate readers about the substance, importance, and power of evolutionary theory. Richard Dawkins's *The Greatest Show on Earth: The Evidence for Evolution* (2009) and *The Magic of Reality* (2011), as well as Jerry Coyne's *Why Evolution Is True* (2009) are particularly good at laying out the facts of evolution in terms nonspecialists can understand.

The stakes couldn't be higher.

American historian and environmentalist Wallace Stegner wrote years ago that humans "are the most dangerous species of life on the planet, and every other species, even the Earth itself, has cause to fear our power to exterminate. But we are also the only species which, when it chooses to do so, will go to great effort to save what it might destroy."[11] That's because ours is the only species whose members realize they are profoundly linked to the rest of the natural world, who believe they have the ability to improve their conditions, and who can cooperate to do profound things that effect positive change. Within our very nature we have the ability to shape the informational and technological environment in ways that can preserve and enhance life, not diminish or destroy it.

CHAPTER 1

THE GREAT CHAIN OF COMMUNICATION

Vast chain of being! which from God began,
Natures ethereal, human, angel, man,
Beast, bird, fish, insect, which no eye can see,
No glass can reach; from infinite to thee . . .

Heaven's whole foundations to their centre nod
And nature trembles to the throne of God.
All this dread order break—for whom? for thee?
Vile worm!—oh madness! pride! impiety!

—Alexander Pope, *Essay on Man* (1734)

In his celebrated work *Essay on Man*, the eighteenth-century British poet Alexander Pope expressed what had become the guiding philosophy of the day—unshakeable belief in a fully ordered and stable world ordained and supervised by God.

The evocative metaphor Pope uses repeatedly—the "Chain of Being"—derives from essentialist conceptions of nature and the universe. Everything fits into a strict hierarchy that descends from the highest possible metaphysical standing: *en perfectissimum*, God. Ranked below God are celestial beings (angels); then humans (sorted men over women, royalty over peasants, masters over slaves); then animals; then plants; then rocks, minerals, and soil. The Chain of Being is true and complete. In this world there is no mutation, no adaptation, no evolution. Whatever changes take place on the earth reflect only the actualization of a predestined order, the projection of an essence. The logic of the chain is self-evident, and, as Pope warns, only a mad, arrogant, and impious worm would dare to challenge the divine authority that rules over it.[1]

Charles Darwin spent his life inspecting and reflecting on the Chain of Being. The astounding conclusions he would eventually draw, both scientific and philosophical, were not entirely original. Early biologists—especially Georges Buffon, Carl Linnaeus, and Jean-Baptiste Lamarck, all of whom influenced Darwin's thinking in one way or another—also invoked the idea of an orderly arrangement of scaled components in their systems of scientific classification. But as they worked, they began to modify the imagery. Instead of a descending chain, the biologists regarded the vast network of relationships that connects living organisms more as a "Tree of Life" that grows upward toward complexity and diversity, not downward toward simplicity and uniformity. God may be present, they thought, but much must also be explained about how the flora and fauna function and change right here on the ground.

Darwin likewise did not readily dismiss the idea of God. The chauvinistic Chain of Being idea even led him to consider at first that indigenous peoples must represent a species that is closer to animals than to white Europeans. Indeed, some of the tribal peoples Darwin encountered on his journey aboard the *Beagle* themselves struggled to mark clear differences between their own people and the animals around them.[2] Slowly, Darwin rejected the idea of an unverifiable deity ruling autocratically from the top of a rigid hierarchy in nature. He expanded the biologists' alternative Tree of Life metaphor in his discussion of natural selection in *On the Origin of Species* and illustrated the new diversifying picture of life graphically.[3]

As he developed the theory of natural selection, Darwin also came to believe that the Tree of Life represents not only direct interconnections among living things—existing roots, trunks, limbs, branches, twigs, and buds—but also the relation of contemporary life-forms to other forms in the past and future. Time, thus, became a primary consideration. He wrote that the Tree of Life "covers the [earth's] surface with its ever branching and beautiful ramifications" in the magnificent present. But it also "fills with its dead broken branches the crust of the earth" (the evolutionary past), while "buds give rise by growth to fresh buds, and these, if vigorous, branch out and overtop on all sides many a feebler branch" (the evolving future).[4] Darwin was imagining the radical idea of deep time.

Darwin's theory of natural selection would render any idea of fixed positions along the traditional Chain of Being obsolete. Biological entities would never again be understood as immovable links. Even rocks and minerals could rightly be considered dynamic parts of the chain because they reveal the nature of environments where life existed before and serve to sustain life now. The hierarchy represented in the Chain of Being had things exactly backward; life springs naturally from the bottom up, not the other way around. The idea of the prescientific Chain of Being seemed like an overly determined implement planted firmly in an undetermined world. The metaphor was wrong. A sturdy ladder or stairway to heaven would have been a more appropriate image.

Darwin and many of his fellow naturalists saw a new image of life emerging. Evolution does not proceed by divine intervention from the top, and it does not develop willy-nilly from the bottom. A much more complex and delicate set of factors and processes is at work. Nature is characterized by interconnectedness, movement, and change with no designer directing the action. Life can be sustained only by the production of interactions that work to the advantage of the organisms involved in the particular contexts they inhabit. In the process, the natural world is being made and remade constantly. But what underlies, facilitates, empowers, and regulates the incessant change?

We suggest an answer to that question in the following pages by giving the chain new meaning and relevance. Living things are linked to each other, yes, but the links don't materialize as solid entities. If life processes can be symbolized fruitfully as the links of a chain, then the chain should not be thought of as a series of domains that in any way freezes the elements into place or constrains their potential. The links can best be understood as flexible spheres of robust connectivity that flow within, between, and among biological agents, unifying all of nature in the process. Only one word accurately describes the ground on which such processes unfold: communication. Organisms survive and flourish in this world because they have the ability to communicate effectively.

EVOLUTION: BORN OF RELIGIOUS CONTROVERSY

As a parent explained to a psychologist at the University of Maryland when asked if she believes in evolution, "I don't know what to believe . . . I just want my child to go to heaven." Her case is typical. Far less than half of the American public believes in evolution, and more than 90 percent believe in a personal god.[5] More people believe in angels, extraterrestrial beings, the devil, heaven, and literal hell than in evolution.[6]

The unwillingness to question the basis of religious belief functions like an analytical blind spot. Charles Darwin and many other nonbelievers of his era endured the same irrationality and resistance that evolutionists face today. The dustup over "intelligent design" is just the latest instance of the tiring "debate."[7] Evolution is as certain as gravity, but many people can't accept the fact that humans, like all the other primates, descended from an apelike ancestor. It challenges the core of their identity, their sense of well-being, and their hopes for going to heaven—the key sticking point. In that respect, not much has changed since *On the Origin of Species* was published more than 150 years ago.

Religion provides a convenient escape from another conclusion that most humans find very uncomfortable: we come from lowly origins, just as Darwin said.[8] Americans practice voluntary ignorance about this issue with special zeal. Only half of the American public believes that humans developed from other animal forms.[9] That should come as no surprise. In the United States, God, not evolutionary processes, has been identified most often by citizens as the "Creator" of human beings in research that dates back decades.[10] Evangelical Protestants are particularly likely to deny the truth of evolution.[11] Americans are not the least enlightened in this regard. Among citizens of populations sampled in a survey conducted by the National Center for Science Education, Turkish people are even less likely to think humans developed from other life-forms.[12] And Turkey is a secular state whose religious culture is moderate by Islamic standards.

Some antievolutionists believe that evolutionary theory means humans descended from apes that resemble the ones that roam the world today. Of course, that is not what Darwin claimed. Humans and today's great apes—bonobos, chimpanzees, gorillas, and orangutans—all descended from a common ancestor. That's why we share so many structural similarities and so much DNA with our primate cousins. As Darwin describes it: "Man descended from some less highly organized form. The grounds upon which this conclusion rests will never be shaken for the close similarity between man and the lower animals in embryonic development, as well as in innumerable points of structure and constitution . . . are facts which cannot be disputed."[13] Humans branched off from the other apes more than five million years ago and began to assume a distinctive anatomy more than one hundred thousand years ago.[14] To argue otherwise distorts the true story of our species.

Throughout the centuries Jewish, Christian, and Muslim clerics have rejected evolution as one way to distance humans from animals and nature. The Book of Genesis makes it clear to Jews and Christians that God created man in his image, not in relation to other life-forms. Islamic creationism is less specific but just as hierarchical. Like the proponents of the traditional Chain of Being confidently proclaimed, all three major religious belief systems continue to assert that humans exist above animals and sit close to God.

Mean-spiritedness often characterizes public attacks made on evolution and evolutionists. Never was this more evident than when the citizens of Dover, Pennsylvania, voted out school board members who had approved including intelligent design in a high school biology course in 2005. Popular religious broadcaster Pat Robertson subsequently warned on his nationally syndicated television program, "If there's a disaster in your area, don't turn to God. You just rejected him in your city . . . if they have future problems in Dover, I recommend they call Charles Darwin. Maybe he can help them."[15] A district court judge later provided exacting legal justification for removing the creationist material from the classroom. Responding to a mountain of evidence presented by experts, the judge concluded that evolutionary theory "represents good science" while intelligent design amounts to nothing more than "a particular version of Christianity."[16] The intelligent design crowd was dealt another blow a year later when the State of Kansas Board of Education voted in new members after the previous group required teachers to strongly criticize the validity of evolutionary theory in the classroom, paving the way for teaching creationism as science.[17]

The popular appeal of religion and the expansion of modern media have given commentators like Pat Robertson and the late Jerry Falwell, founder of the Moral Majority, great opportunities to use God for ungodly purposes. From Father Charles Coughlin's radio broadcasts attacking Communists and Jews in the 1930s to Glenn Beck's incendiary messages on radio, television, and the Internet today, the electronic media have greatly intensified the influence of religious zealots. In the wake of the 2001 attacks on New York City and the Pentagon, Osama bin Laden became a media celebrity with a global platform. Thanks to the Internet, American-born Yemeni imam Anwar al-Awlaki emerged as one of Islam's most recognizable and influential jihadists before he was killed in a drone attack.

JUST A THEORY?

A sleight of hand commonly employed by antievolutionists is to argue that evolution is "just a theory." This manipulation of language confuses imprecise popular meanings of theory with precise scientific understandings. In popular discourse, *theory* has strong negative connotations. Theory is nothing more than a point of view that cannot be properly verified or supported—often mere speculation or conjecture. In the State of Georgia, for instance, a sticker is placed on the cover of biology textbooks stating "Evolution is a theory, not a fact."[18] Theories are considered particularly dangerous when they challenge widely accepted beliefs and practices.

For scientists, *theory* refers to something far more specific and respectable. An established scientific theory is a well-supported position on an issue that emerges gradually from the rigorous testing of conditions and relationships in the empirical world. Any thesis must meet the standard of falsifiability; you have to be able to test the theory. Propositions that are not supported by evidence or that cannot be tested empirically—like claims that biological adaptations in the natural world are caused by the hand of God, or that life began in the Garden of Eden six thousand years ago—do not count. Popular appeals don't count either. The top-selling "nonfiction" book *Heaven Is for Real* (2010) was based on the story of a boy's encounter with a blue-eyed Jesus and flock of adoring angels when the child underwent general anesthesia during an emergency appendectomy at age three.[19] These whimsical assertions and many others like them fall victim to the fallacy of *argumentum ad consequentiam*—literally, "argument to the consequences": an outcome is true or false based on how much the proponent likes or dislikes the consequences.

Although the implications of genuine research and theorizing can be shocking—as the explanation of human evolution certainly has been for some people—scientific theories develop very conservatively. They will not be accepted by ethical scientists unless the method for collecting and analyzing the relevant data conforms to strict procedures and limitations. Conclusions must be carefully drawn according to the quality of evidence and cogency of reasoning used to explain the evidence.

Evolution is a theory in this scientific sense. All forms of creationism are not. Ironically, the popular interpretation of scientific theory as mere speculation applies to claims made by religionists, not evolutionists. And it's bad speculation at that. Arguing "You can't prove that God doesn't exist," for example, does not mean that he does. This is the *ad ignorantium* fallacy: a claim is believed to be true because it hasn't been disproved. The idea that the human body decays after death but that the soul flies up to heaven is scientific nonsense. A baby born with two faces does not mean God has produced a reincarnated Hindu goddess. The theory that dinosaurs existed in the distant past and that traces of their existence persist in life-forms today, however, is well supported by fossil and DNA evidence. The dinosaur theory passes the tests of scientific integrity with flying colors. Literally. The structural features and DNA of certain ancient dinosaurs are present today in the brightly colored birds that soar over our heads![20]

Evolutionary geneticist Jerry Coyne takes the argument about the legitimacy of evolutionary theory one step further: "Evolution is far more than a 'theory,'" he says. "Evolution is a fact."[21] He points out that "a theory becomes a fact (or a 'truth') when so much evidence has been accumulated in its favor—and there is no evidence against it—that virtually all reasonable people will accept it."[22] Such should be the case with evolution. One might "believe" in God, but there is no need to "believe" in evolution. Still, despite scientific advances that demonstrate its historical factuality and its theoretical truthfulness, evolutionary theory remains painfully problematic for most people around the world, even those who act sensibly most of the time.

Coyne explains, "We are apes descended from other apes, and our closest cousin is the chimpanzee, whose ancestors diverged from our own several million years ago in Africa. These are indisputable facts."[23] But Coyne believes the fact that we have evolved from apes is not the main stumbling block to accepting the truth of evolution. Instead, he says, it is "the emotional consequences of facing that fact . . . [T]o these folks, evolution raises such profound questions of purpose, morality, and meaning that they just can't accept it no matter how much evidence they see."[24] Confronting the reality of evolution is no easy task for persons who have given their soul, or pretended to give their soul, to a supernatural being, and their time and money to the cultural institutions associated with it.

To accept that evolution is true calls into question the validity of the commonly accepted myths, rituals, and other forms of religious and cultural reinforcement that pretend to explain the origin and meaning of life. Granting the plausibility of evolution gives credence to the atheist and the agnostic, a compromise most religious people are not willing to make. Yet sometimes the empirical evidence is so compelling and the philosophical stakes so high that boldly and publicly stating a theory as factual and true seems necessary. This was the case with Stephen Hawking's recent declaration that "spontaneous creation is the reason there is something rather than nothing, why the universe exists. Why we exist. It is not necessary to invoke God to light the blue touch paper and set the Universe going."[25] In an earlier era, the same credit can be given to Charles Darwin for his courageous assertion that evolution is factually indisputable.

DARWIN LOVES YOU

A clever bumper sticker appropriates and ridicules a familiar Christian refrain: "Darwin Loves You." Meant to be comical and ironic, the message is actually true. Charles Darwin was a gentle and loving person, a caring husband and father, and a supporter of the Anglican Church his wife and children attended near their Down House estate outside London. His professional manner was equally dignified. Reading *On the Origin of Species*, we find an exemplary scholar—careful, respectful, and modest. After meticulously laying out the empirical conditions that led to his theoretical claims, he always specified what could prove his theory wrong. That's excellent science and a clear indication that he was a person of exceptional moral and intellectual integrity.

Moreover, Darwin himself never scorned religion. The young Darwin came from a family of churchgoers. His wife was a devout Christian, and he was buried at Westminster Abbey. As he set out to sea from Plymouth, England, at age twenty-two for his five-year trip around the world, Darwin held no grudge against God—to the contrary. He took a Bible with him on the voyage and considered himself to be a Christian.

Serious doubts about the existence of God, however, started to eat away at him during the journey. The conclusions he was drawing as a naturalist over the months and years contradicted the assertions of religious dogma. One personal incident stood out. Darwin observed Christians as slave owners for the first time when the *Beagle* docked in the state of Bahia, Brazil, in 1836. He was greatly troubled by the harsh realities of slavery and by the Portuguese Catholics' use of biblical scripture to justify the practice.[26] Darwin became a dedicated abolitionist. His hatred of slavery coincided with his thesis about the common descent of all human beings, and it probably shaped his overall view of evolution more than is commonly acknowledged.[27] Charles Darwin believed deeply in freedom, justice, and the human potential. These qualities of Darwin the man and Darwin the scientist are often misunderstood or misconstrued.

Strong criticism comes at Darwin and evolutionary theory not only from the religious right but, for very different reasons, also from the political left. Two recurring charges from the left are racism and sexism. Darwin held and expressed racist and sexist opinions in his writing, as was common at the time. But what does evolutionary theory actually say about racial differences? Did Darwin believe evolutionary processes have forever determined our gender roles? Does evolutionary theory feed racial and gender prejudice?

RACE

The impression that Darwinian evolutionary theory is racist springs from the "survival of the fittest" idea, wherein it is widely perceived that some "races" are claimed to be more likely to survive and to be civilized. Indeed, Darwin did observe that some groups seem to fare better than others. Darwin also used the terms *savage* and *barbarian* to refer to some of the indigenous peoples he encountered in his travels. He wrote things like "negroes" are excellent musicians "although in their native countries they rarely practice anything we should consider music."[28] Crude as those words, characterizations, and judgments may seem today, they reflect the common vocabulary and thinking of the time—a period that long predates the changed meanings of language and cultural sensitivities expected today. To put matters into proper context, blacks and women were not even considered to be "persons" under terms spelled out by the most enlightened political document of the time and a model for governments throughout the world today—the United States Constitution.[29]

The more substantial discussion of "survival of the fittest" (a term that originates with the English social philosopher Herbert Spencer, not with Darwin) centers on the implications of evolutionary theory concerning the relative ability of various groups to survive in a world of exploding population growth and declining food resources. The imagined perils of evolution are frequently associated with the grim hopes for human survival forecast in Thomas Malthus's *Essay on Population*, published a few years before Darwin was born.[30] The solution Malthus and others proposed to curb competition over scarce resources—voluntary population control through abstinence—reveals just how serious the problem of overpopulation seemed then. Though it certainly was never Malthus's intention, the extreme competition described in his writing has been used by nefarious political authorities to justify policies of social engineering and ethnic cleansing in the name of nation and race.

Just the word *evolution* repels some scholars in the humanities and social sciences. The reluctance to apply the established tenets of evolutionary theory to phenomena in the social sciences stems in large part from overheated discussions in the 1970s about sociobiology, a perspective that generalizes from the behavior of lower-order species to help explain why humans act the way they do. For some, sociobiology represented an assault on disadvantaged members of society. As pointed out by the editors of the journal *Nature*, many academics on the political left in the 1970s categorically rejected "biological explanations for phenomena such as gender roles, religions, homosexuality, and xenophobia largely because they feared such explanations would be used to justify continuation of existing inequalities on genetic grounds."[31] Those fears still linger in some quarters.

Like all thoughtful scientists of his time, Darwin took Malthus's influential essay into serious consideration. He reasoned that if Malthus's dire calculations were true for men and women in nineteenth-century industrial societies, then they also had to be valid in the wild, even more so. Animals procreate as much as possible; each generation unconsciously acts in a way that geometrically increases the number of its offspring. They don't have the option of using technology to increase their food supply or to willingly control the size of their populations. Life in the wild pits population growth brought about by ceaseless procreation against population reduction caused by ever-present predation.

Darwin understood that the biological universe is forever engaged in a mighty struggle for survival, with many casualties. Yet he made it clear that his theory does not support any kind of doomsday scenario that privileges one group of people over another, no matter how desperate they become. We may have diversified as a species over the millennia, Darwin said, but we all belong to the human family. He could not conclude, for example, that the indigenous natives he met on the *Beagle*'s swing through the Southern Hemisphere were different from Englishmen. By contrast, some other European naturalists of the same era considered Africans to be of another species. A fellow British naturalist, Louis Frazer, even shot a young native boy, thinking he was a monkey.[32]

The evidence was clear to Darwin: because the human races are so similar, they must have descended from a common progenitor.[33] For that reason "there is no justification for placing man in a distinct order," an opinion made explicit in Darwin's *Descent of Man* and trumpeted by one of the great early defenders of early evolutionary thinking, biologist Thomas Huxley, known as "Darwin's Bulldog."[34] The matter of racial differences nonetheless continues to provoke controversy and requires a careful interpretation of human history. It's astonishing to see how Darwin's groundbreaking insights—prompted only by observing nature, peering through the lens of a tiny microscope, tending to domestic animals and plants, and corresponding by overland mail with colleagues—have been so roundly confirmed by the vastly improved scientific evidence generated by modern research techniques.

As *Homo sapiens* dispersed globally over the past fifty thousand to sixty thousand years, various groups settled in relative isolation from each other. Morphological contrasts and diverse social and cultural traits—languages, music, religions, and ways of perceiving time and space among other characteristics—emerged among distributed populations over long expanses of time.[35] Yet any single person's race is not easily discerned. We are black, brown, yellow, and white only by degree. "Race" is more a convenient political category than a precise biological type. Global trade, travel, and immigration have produced a degree of genetic convergence that makes racial identity based on continental ancestry problematic. This has led to another controversy linked to evolution. Ancestry tests using DNA have become popular as people attempt to trace their racial and ethnic heritage, sometimes with shocking and personally troubling results. For example, North Americans who have always considered themselves white may discover they descend not only from Ireland but also from East Asia. Some African Americans have found out how little African blood they really have. And Adolf Hitler would surely have been distressed to learn his DNA revealed the possibility of Jewish and African ancestry.[36]

Some racists apparently believe their views can be justified by evolutionary theory. They are wrong. Charles Darwin had it right when it came to race. We descend from common origins, yet our species is variable. Common wisdom and decency dictate that we don't make too much of the differences. Every good-spirited person judges others by the content of their character, not by the color of their skin. In order to protect the rights of racial and ethnic minorities, this fundamental principle has been codified into civil rights laws in modern societies everywhere.

SEX AND GENDER

Like religion and race, discussions about sex and gender in human evolution often stoke controversy. Evolution is frequently considered to be prejudicial toward women, and Darwin is often chastised for the language he used and some of the conclusions he reached about sex and gender. Is evolutionary theory biased against women? What did Darwin actually say and what does it mean today? To shed some light on these issues, we quote and paraphrase Darwin in the next few paragraphs from passages that appear in a section of *The Descent of Man* titled "Difference in the mental powers of the two sexes."[37]

Darwin argues that just as the males of other species have developed special physical characteristics that attract females for mating—brightly colored feathers, for example, or an impressive rack of antlers—man's "struggle for life" against nature and his "contests for wives" with competing males have also had the cumulative effect of producing pronounced physical differences between the sexes. Males and females gradually became dimorphic—of differing physical forms.

Those aren't the only dissimilarities. Men and women, he said, also differ greatly in "mental disposition." For instance, men "delight in competition," according to Darwin. They are courageous, perseverant, determined, and ambitious to the point of selfishness. But the "chief distinction in intellectual powers" between the sexes lies in what men and women have been able to achieve in life. Men attain a "higher level of eminence" in whatever they try to do than do women—whether it requires "deep thought, imagination, reasoning, or merely the use of the senses and hands." One reason for their success is that men are more patient; it's part of their "genius."

In Darwin's estimation, women also have useful qualities. They are more tender, less selfish, more intuitive and perceptive, and less competitive. These characteristics, he reasoned, derive mainly from giving birth and taking care of infants and children. But he also believed that women have less "mental power" and are simply not as accomplished as men—not as "eminent" in any professional endeavor. Comparing the sexes, Darwin concluded that "man has ultimately become superior to woman."

"Superior" is quite the harsh assessment. Have men and women evolved this long only to become the stereotypes Darwin describes? In Darwin's day, role expectations between men and women were more sharply differentiated than they are today. But to one degree or another, then and now, the gendered division of labor can be seen universally.[38] Evolutionary history helps explain how things got that way: the division of labor between the sexes developed and spread among social groups because it has proven to be evolutionarily advantageous. Sexual dimorphism and gendered behavior helped foster particular kinds of cooperative social interaction that became expedient for confronting environmental threats. The essential division of labor between early *Homo sapiens*—men hunted large game while women gathered small game and planted and harvested food—may have even allowed our species to survive and thrive while the less gender-differentiated *Homo neanderthalensis* competitors went extinct.[39]

Does evolutionary theory predict that the gender statuses and roles present in today's societies will remain fixed? Of course not. *Evolution* is another word for *change*, and change is not determined by biological difference or cultural history. The social movement toward achieving greater gender equality has itself become a survival strategy for modern societies. In the United States alone, women outnumber men as college graduates. Half of new doctors are women. Women's pay has increased, though it still lags behind that of men. Women's roles and rights are definitely improving overall.[40] The Internet has been a tremendous boon for gender equality. Moreover, the gendered distribution of social influence doesn't favor only males. Females throughout the animal world control sexual reproduction because they select their sexual partners.[41] Males pursue; females choose.

As a general rule, stereotypical male physical and behavioral characteristics became dominant because they are desired by females.[42] Over time, female choice of mating partners creates striking physical differences between the sexes. Males generally became bigger, more colorful, more ornamented, and more physically active. In the few species among which the reproductive strategy is reversed, the females have become more equal in size and more brightly colored and ornamented than the males.[43] Males' conspicuous aggressiveness should not be considered a sign of superiority or character flaw. It represents a desperate and dangerous drive to be chosen for the biggest game in town—sexual reproduction.

Darwin and evolutionary theory should not be blamed for the inequality that pervades contemporary gender relations. The theory reveals something important about *how* sex-role differences and injustices in our species came about; it certainly didn't create them. Darwin was insightfully reporting nature's cruelty, not endorsing it. While his blunt assessments and choice of language may provoke controversy today, his concerns were purely scientific.

Yet Darwin was personally troubled by what he saw. Darwin made it clear in *The Descent of Man* that he considered the lack of equality between men and women to be an important and addressable social problem. When he claimed that males are "superior" to females, he really meant that they are evolutionarily advantaged. Things didn't start out like that. Our hominid ancestors first emerged as sexually androgynous creatures with undifferentiated biological and social roles. Since then, no human being has ever been born purely male or female. The essential unity of descent is inscribed into our very bodies. As Darwin wondered, why else would men have nipples?[44]

Behavioral differences we think of as gender stereotypes began to evolve once our human ancestors started to reproduce sexually some six to ten million years ago. Mating competition eventually made males physically larger than females and less vulnerable to attack. The instinct to protect themselves and their progeny forced males to excel outside their immediate surroundings, a trait that persists in some quarters today. By the time Darwin was writing in the nineteenth century, inequality between the sexes in cultures all over the world had become so apparent that he must have concluded that the differences could only have been created by incremental role segregation accumulating over countless generations.

Who ultimately holds the power in human sexual relations? The popular young American actress Megan Fox voices a blunt opinion: "Women hold the power because we have the vaginas. If you're in a heterosexual relationship and you're a female, you win."[45] Her comment reflects a clear middle-class Western bias, but she has a point. Females throughout nature choose males for sexual reproduction. But nature's norm doesn't apply to humans in such a straightforward way. Men typically assert greater influence than women over the terms of sexual life and reproduction, especially in deeply traditional societies. Many marriages are arranged by male-dominant families. Dowries are offered. Girls are sold. Women don't have equal access to education, voting, property ownership, even simple freedom of movement. Covering females from head to toe serves the purposes of men. The ultimate control occurs when girls are murdered by the men in their families for violating social and sexual restrictions. By contrast, heterosexual interaction in most modern societies doesn't conform strictly to tradition, nor does it necessarily lead to a lasting commitment. Men still prevail in most areas of public life in modern societies, but both males and females influence sexual interaction and reproduction.

Sheer muscle power has dwindled as an economic resource as machines have replaced many kinds of manual labor. Skilled industrial trades have given way to less gender-specific expertise required of workers in today's information-based economy. Gender disparities in most all aspects of life have been reduced significantly in modern cultures. Economic, political, and cultural development in the modern world *depends* on cultivating greater opportunities for girls and women, especially where the rights and opportunities for women lag behind.[46] Darwin was convinced that gender inequality could be addressed by intentionally changing social behavior—in effect by training women to act out males' roles, who would then pass the acquired traits along to their daughters.[47] He even spoke of a kind of affirmative action strategy when he observed that plants and animals "placed in a new country amongst new competitors" require some kind of environmental modification in order to adapt and survive.[48]

CHANGE IS (ALWAYS) COMING

When Charles Darwin was alive, the idea of "culture" was not an established scientific framework available to researchers for interpreting the natural and social world.[49] Biological factors dominated early discussions about evolution. Scientists now have a much more precise and nuanced idea of how evolutionary trajectories have been influenced by regional conditions and how culture shapes consciousness and behavior. People fear change in theory and in practice, and the two domains are related. Human beings naturally resist changing the beliefs, affiliations, and practices that organize their daily activities in ways that provide security and meaning to their lives. But change is inevitable and is speeding up. Culture is the space in which adaptive retooling takes place. Social communication is the means by which it is accomplished. The great chain of communication passes through every structure and process of life on the earth, making evolution a unified theory. How communication became such a powerful force in our species' struggle for survival and quest for progress will be explored in the chapters that lie ahead.

CHAPTER 2

COMMUNICATING TO SURVIVE

When British Labor Party politician Jack Straw remarked in 2006 that Muslim women in the United Kingdom should not wear veils covering their entire face in public, he added fuel to a fire that continues to burn in Europe and elsewhere. The Netherlands banned the full veil months later, and Belgium passed similar legislation. Six years after the French government banned wearing headscarves and other forms of religious apparel and jewelry in public schools, the National Assembly and Senate passed legislation to outlaw wearing the full veil by a near-unanimous vote. Majorities of the public in Germany (71 percent), Great Britain (62 percent), and Spain (59 percent) said they would support a similar ban in their own countries.[1] Syria has banned scarves at universities that leave only an opening for the eyes, and a debate over wearing headscarves in public has raged for many years in Turkey.

Some object to the full veil for security reasons. Others worry about religious culture intruding into secular life. For his part, Jack Straw insisted that he only wanted to improve the quality of public communication in Great Britain. Like many people, Straw found it alienating and unproductive to deal with individuals who shut down a vital source of information for person-to-person interaction—the stream of nonverbal cues given off by facial expressions. Politics aside, he's right. We communicate to survive. Intercellular messages flow inside our bodies to increase biological fitness. We interact with others to navigate earthly environments safely. Having access to information and the ability to exchange messages through multiple channels sustains and enhances life.

Charles Darwin was also disturbed by the way veils inhibit communication. Focusing his analysis on the moral implications of veiling for communities, he considered the practice to be among humankind's "strangest customs and superstitions," a cultural behavior that stands "in complete opposition to the welfare and happiness of mankind." Within the Muslim tradition, the eyes of men and women who are not related are not supposed to meet. Attempting to explain the roots and consequences of this behavior, Darwin concluded that belief systems and cultural conduct are "constantly inculcated during the early years of life, whilst the brain is impressible, appear to acquire almost the nature of an instinct; and the very essence of an instinct is that it is followed independently of reason." He argued that the power of such deeply ingrained habits is revealed, for example, by the "severe remorse" a Hindu would experience for eating "unclean food" or the "horror" that would be felt by "a Mahometan [Muslim] woman who exposes her face."[2]

Animal communication serves as a pillar of biology but is rarely recognized as such. Cells and multicellular organisms can only be properly understood in terms of their mechanisms of interaction. The complex world we live in today evolved from the most basic form of biological interactivity—cell-to-cell communication that takes place within a living organism. Single-cell organisms grew to become multicellular animals and plants because some life-forms, acting over vast expanses of time, reacted favorably to changes taking place in their physical environments. Propagative disaster was avoided because cells interact in ways that forestall a rate of reproduction that would overwhelm the system.[3] Some organisms became more elaborate by further adapting to environmental changes and passing on their genes—parcels of information that reflect specific hereditary traits transmitted by plants and animals to their progeny.

Genes reproduce and communicate information with extraordinary accuracy, but they don't always make perfect copies. The irregular gene pool is filtered through endless generations that further mutate and produce diverse biological forms. While the results of this process aren't random, they aren't predetermined either. Certain genes become favored over others in the rough-and-tumble course of natural and sexual selection. Our bodies, like all living things, "are machines programmed by genes that have survived," according to Richard Dawkins.[4] We humans are the lucky ones. Countless other biological forms—from simple bacteria and tiny beetles to majestic dinosaurs; woolly mammoths; and our hominid cousins, the Neanderthals—are buried deep in the boneyard of evolutionary history.

Biological evolution arises from parallel communication processes—intra-organic cellular interaction and the transfer of genetic information from one organism to another. Life-forms survive because they mutate and change. Genes that positively influence the host organism's prospects for survival dominate the gene pool.[5] As Charles Darwin recognized long before the discovery of genes and DNA, organisms that behave in ways that aid in their survival without understanding what they're doing are engaged in what he termed "unconscious selection."[6] While genes perform as the basic unit of natural selection, they don't do it alone. Genes turn other genes on and function in clusters of DNA, the molecular component of heredity. In what became an enormous leap forward for evolutionary science, the genome—representing the totality of an organism's DNA—provides remarkable insights into our physical development, cultural histories, and mental maps.

WE'RE EMOTING, COMMUNICATING ANIMALS

We advertise our physical and emotional needs and desires through language, nonverbal communication, music, and countless other forms of expression. This behavior is not exclusive to humans. Charles Darwin wrote extensively about how dogs, including his own canines, send messages of need, affection, dominance, territory, sadness, and fear. Birds sing, chirp, strut, and display their feathers to communicate their intentions. Chimpanzees grin, laugh, and chastise each other. Frogs croak out territorial claims and call potential mating partners. Crickets and other insects pierce the air with sexual signals and messages of contentment, then go silent when danger lurks. Lions roar to scare away intruders and call the pride to reunite. Lightning bugs flash intermittent signals of availability. Giraffes position their necks in ways that express fear, panic, anger, and submission. African wild dogs lick each other and run around together in a frenzy to create a prehunting mood. Bats and dolphins interpret echoes from sounds they emit to navigate risky environments and find food. Even many species of fish make vibrating noises when competing for food or sex. The search for food and sex, and the need to steer clear of predators compel all kinds of creatures to communicate incessantly.

Does some psychological commonality lie at the root of these forms of expression connecting us to the rest of the animal kingdom? From personal observation and the reports of many of his colleagues around the world, Darwin concluded that higher-order mammals express a wide range of emotions, including anger, love, grief, and shame. According to him, primates "all have the same senses, intuitions, and sensations—similar passions, affects, and emotions, even the more complex ones such as jealousy, suspicion, emulation, gratitude, and magnanimity."[7] For instance, Darwin noted how monkeys, when "pleased, utter reiterated sounds, clearly analogous to our laughter, often accompanied by vibratory movements of their jaws or lips, with the corners of the mouth drawn backwards and upwards, by the wrinkling of cheeks, even by the brightening of eyes."[8] Monkeys use other facial expressions and body postures to intimidate rivals. Chimpanzees depend in part on subtle facial expressions—the arch of an eyebrow, the curve of a lip—to understand each other.[9] Talapoin monkeys of West Africa rely on finely textured vocal signals rather than on nonverbal signals to communicate with each other because their habitat—thick treetops—obscures visual contact.

We feel close to animals when they display emotions that resemble our own. People ascribe humanlike qualities to animals, especially to domesticated dogs, by interpreting their feelings, often in self-serving ways—"You missed me, didn't you Poochie? I love you too!" This tendency is evident in a letter published in the *San Jose Mercury News* that describes how a dog belonging to an elderly couple had been run over by a car:

> A dog died on Squiredell Drive the other day. His name was Tucker and he was one of those special dogs that leaves us too soon. . . . He was such a special, loving dog that he wanted all his people friends to take his goodness and spread it around. Tucker would not want his death to foster anger and blame. He would not want to be remembered that way. He was a good dog and his dying breath was not about asking why, but telling all of us how grateful he was for letting him touch their lives.

Emotionally expressive chimpanzees can be so humanlike that some people adopt them as pets. Probably the most famous chimp owner was Michael Jackson, who raved about how much he and the children at his Neverland Ranch in California enjoyed sharing meals with Bubbles and Max. Jackson claimed his chimps were extremely friendly and polite. Certain inconveniences cropped up, however. Although Bubbles wore diapers and had been trained to use MJ's personal toilet, domestic help at the ranch complained that the chimp frequently threw fecal matter against the walls. Undaunted, the late King of Pop dressed himself and Bubbles in matching outfits, hired a bodyguard to keep his companion safe, and took his primate pal along with him to recording sessions. But people who adopt or work with chimpanzees need to be protected. Chimpanzees can become aggressive and extremely dangerous even when raised from birth in the homes of human families.

Emotional expression is survival signaling that typically commences at the very beginning of life. Feelings of physiological need, discomfort, disorientation, and insecurity cause human babies to signal caregivers by means of intense vocal and bodily expressions—the first communicative acts a newborn child performs. Parents sense the child's emotional state and respond, positively reinforcing the expressive act. Infants and children create other kinds of expressive displays by mimicking other people. Their expressive creativity arises instinctively when they spontaneously draw, mold clay, and fold paper, among other artistic endeavors.[10] Human bodies evolved in ways that facilitate a wide range of expression. Well-developed facial musculature gives humans the most malleable countenance of all the primates.[11]

From the naked human body to the most sophisticated communications technologies, people use the expressive media available to them to transform their inner thoughts and feelings into direct physical signals and symbolic representations. Expression and symbolic creativity are not "add-on" features of modern life. Their platforms, portals, and prospects are central to human existence, the significance of which has increased markedly with the roughly simultaneous arrival and rapid expansion of media channels—symbolic resources delivered by the culture industries, the Internet, and personal communications technologies. Digital technologies greatly increase the speed, efficiency, and intimacy with which high-resolution visual imagery and crystal-clear sound are delivered. Sony marketed its hugely successful PlayStation®, for example, as "the emotion engine" because of the unit's ability to move internal data at vastly increased speed, thereby escalating players' emotional involvement with the games. Speed stimulates the pleasures of expression. When computer users upgrade to faster operating systems or add more rapid access memory (RAM), they are demanding quicker machine responses to their creative impulses and more space with which to exercise their expressive potential.

These interacting fields of force mark the uniqueness, importance, and impact of the Communication Age. But the roots and significance of sentiment predate even the dawn of human history. Emotional expression was necessary for the progenitors of modern humans to forge bonds of increased social solidarity, to create stable local-group structures, and to form the first true cultures.[12] Gestures and meaningful vocalizations gave early primates an evolutionary advantage that created consequences for modern humans. The constant pressure of the environment against the organism provoked creative adaptive responses that shaped the history of our species. As American sociologist Jonathan Turner puts it, "Not only did [natural] selection produce an animal capable of controlling emotions and using them to communicate meaning, selection also worked to create an animal able to generate a wide array of emotional states. The complexity of human thought, social interaction, organization, and cultural life is not possible without this ability."[13]

SENDING SIGNALS

Emotional expression functioned as a key adaptive mechanism for our evolutionary ancestors and continues to do so for humans and other primates today. Of course, the functionality of communication is not limited to emotional expression, important as it is. Communication orders life. The basic need to share information, even when doing so is not consciously intended by the message sender, also motivates communicative action. The classic example is the way honeybees excitedly signal each other when they discover food. The nature of the food, the direction of the food source, and its distance from the hive are communicated to the hive by the returning bees using "waggle dance" language.[14] The colony then responds to news of available food by further transmitting the information internally.[15]

Africa's vervet monkeys have an even more complex signaling system. The monkeys vocalize different alarm calls depending on what kind of predator they are trying to avoid. Leopard calls cause vervets to run up into trees. Eagle calls provoke them to look skyward. Snake calls prompt the monkeys to stand on two legs and survey the ground around them.[16] Any noise made as a warning carries a risk to the monkeys because it attracts the attention of predators. It is not certain whether they intend to warn each other or are just trying to discourage an attack, but the benefits must offset the costs of audible signaling, or the behavior wouldn't have evolved.[17]

Evolution turns a functional action that is characteristic of a species into a communications signal. Behavior is information that is selected, but an accurate interpretation of the signal must take place for the species to select and survive. Honeybee movements, for instance, fit into a broad pattern of communication that begins by indicating closeness to or distance from a food source. A longer flight to the food requires the bee to spend more energy, so the dance is slow; the opposite happens when the food source is close. The act of dancing also emits the particular odor of the pollen, another clue used by the other bees to mount a successful trip. After locating food, the foraging bees commit the information to memory and call on their stored knowledge for subsequent trips, adding another dimension to the insects' communication system.[18] The system ultimately combines "public information" (signals given off to others in the dance) and "private information" (the bees' information recall).

The same kind of analogical communication occurs even when individuals of different species interact. For example, a dog will look at his master's finger pointing in a specific direction and will run without hesitation in that direction when a ball is thrown—even if the ball is an imaginary one. Even if no ball is thrown, the dog often returns to the place where he thinks the ball must have gone and continues the search. Dogs remember the game and beg to play it again and again.

Physical activity thus becomes a communicative "movement-signal"—whether it's the jittery dance of honeybees indicating the location of pollen or a man pretending to throw a ball. The signals that animals send tend to be redundant and progressive. They indicate the appropriate level of intensity by repeating and reinforcing the message. The communicative sequence typically moves from a slow to a more frenetic intensity, especially in competitive situations.

A rhesus monkey, for example, begins his territorial display with a signal of low intensity: he stares at his rival. Next, the monkey stands up. Then he thumps the ground with his hands while simultaneously displaying his teeth in a threatening manner. If the rival does not back off, the monkey attacks. Green herons similarly inflate signals when challenged. They first ruffle their crest and twitch their tail. If that fails to chase the competitor away, the heron inflates the signal by expanding his crest and bringing other feathers to full size.[19] The Sunbittern, a tropical bird of Central and South America, opens its wings and tilts them forward to appear larger and to reveal eyespots—markings that resemble enormous eyes—to ward off predators. Sunbittern chicks begin to exhibit this behavior at three weeks of age.

Charles Darwin noted that the "erection of dermal appendages" such as the "involuntary erection of hairs and feathers" on animals (including the mane of the lion and the ruffling of birds' feathers) also sends strong signals. These communicative "actions," as Darwin termed them, "make the animal appear larger and more frightful to its enemies or rivals."[20] For example, male Nyalas, a Southern African antelope, walk stiffly and raise the hackles on their necks to show dominance. The Visayan Warty Pig, native to the Philippines, and the Chacoan peccary, a South American wild pig, raise their stubbly hair when frightened. Dogs stand tall on their toes with hackles raised, ears erect, tail held stiffly up, eyes transfixed on the target. The dog escalates the threat by retracting the upper lip to "unsheathe the front incisors and large canines." This snarling action is often followed by a menacing low growl in immediate preparation for attack.[21] When animals don't like their chances in a fight, they send the opposite message weighted in degree by the intensity of the threat. We've all seen dogs assume the position of presumed defeat and a plea for mercy. Such submissive behavior reflects what Darwin called the principle of antithesis in animal communication.[22]

THE ORIGINS OF HUMAN COMMUNICATION

Gestures similar to those made by some of today's great apes probably emerged as the first forms of human communication. The vocal modality developed much later, and external forms of representation, including petroglyphs, pictographs, and writing, came into being long after that. The chief proponent of this well-supported theory on the origins of human communication, Michael Tomasello, believes that pointing and pantomiming (mimicking or iconic gesturing) are "the original fonts from which the richness and complexities of human communication and language have flowed."[23]

Our ancestors found ways to communicate effectively because working together improved individual and group prospects for survival. As science journalist Nicholas Wade explains, "Our ability to function as team players in coordinated groups enabled our species to achieve world dominance, replacing other hominids and many other species along the way."[24]

Cooperative communication that facilitated team play grew from a host of gradually expanding patterns of social interdependence. A vital process had been set in motion: the instinct to collaborate gave rise to the first forms of communication, which led to more sophisticated collaborative efforts and more complex ways of communicating in an unending coevolutionary spiral.[25]

Pointing and pantomiming functioned as the primordial antecedents for the development of highly sophisticated, conventionalized sign systems and the corresponding growth of human civilizations. But in some ways, today's societies haven't changed that much from the social worlds our early ancestors created even before the first hominids split off from African apes five to seven million years ago. Communication ability also supports the viability of group life for other species today, especially for primates. Chimpanzees groom each other, develop friendship circles, show sympathy for victims, make up after fights, and work together to chase monkeys and outsider chimps in the wild. They will perform altruistic acts for each other and for humans.[26] Thus the gene for cooperation is likely to have been passed down from our common ancestors. While the size and configuration of primate groups differ greatly, social communication helps all primate species stave off predators, create and sustain relationships and living spaces, protect food sources, and keep enemies at bay.[27] Because communication is good for the community, it is good for individuals who make up the community.

According to Tomasello, human cooperative communication is defined by five characteristics: (1) norms of cooperation; (2) shared goals and communicative intentions; (3) joint attention and common ground; (4) cooperative reasoning; and (5) communicative conventions.[28] Members of a group must feel meaningfully connected to one another for communication to be effective. Individuals identify collective goals and develop norms and conventions for behavior because they believe they belong together. The common ground they share is psychological as much as it is territorial. Cooperative communication and cultural development require long-term interpersonal trust.

Communication ability transforms the biological being into the social self and gradually imbues the human organism with consciousness. As communicative modes and codes evolved, humans increasingly saw themselves reflected in the responses of others, leading them to more fully recognize their capacity to affect the actions of others and to be affected by them. We change the circumstances of our individual and collective lives through intentional acts expressed in the dynamics of social communication.

Humans became excellent communicators and cooperators in part because of physiological adaptations that took place over vast expanses of time. The evolutionary trajectory changed radically when our hominid ancestors began to stand up and walk. That unique behavioral characteristic—bipedalism—may be the first distinctively human attribute to evolve and, in many respects, is the most important.[29]

The emergence of bipedalism traces back to the hominid species *Australopithecus afarensis*, which, according to fossils found in the region, apparently lived in northeastern Africa three to four million years ago. The well-known fossilized skeletal remains of "Lucy" and "Selam" reveal that individuals in this species developed adaptations for walking upright while retaining long arms for swinging in trees to escape predators.[30] The combination of ground living and tree-climbing ability provided evolutionary advantages for a species that was in transition from ape to human.

Moving around on two legs may have made Lucy, Selam, and their relatives more vulnerable to fast-moving predators at first, but over the long run, the advantages of bipedalism outweighed the disadvantages. Bipedalism made it possible for our ancestors to see predators and threaten competitors, wade through water, reach for low-hanging fruit, and tolerate Africa's heat and aridity because walking upright greatly conserves energy, reducing the cost of movement.[31] Running on two legs eventually allowed our ancestors to blend speed of locomotion with endurance in pursuit of protein-rich food—animal flesh—which would help them develop bigger brains and stimulate more advanced cognition.[32] Freeing up the hands gave hominids the ability to carry food, even over long distances. It became possible for them to handle grasses, wood, and stone to fashion primitive tools and to eventually broaden the range of expression. Gesturing, cave painting, body decorating, making simple musical instruments, and writing all developed in part as by-products of bipedalism.

THE TOOLMAKERS

Long before early humans began to decorate their bodies and habitats, they invented crude tools for scavenging, hunting, and eating. The very first tools were probably made of highly malleable materials—twigs, bamboo, and animal parts. We'll never know for sure, because these substances turn into dust with the passing of time. Simple stone-flake tools, on the other hand, preserve well. Because these tools are typically found at archaeological sites that also contain animal bones, researchers conclude they were used for food preparation—cutting up carcasses, extracting marrow from bones, and chopping plants into edible pieces.[33] Learning how to make stone tools required the ability to pass information from one individual to another—a survival-related motivation for developing communication ability and one of the first instances of cultural transmission.

The most important discoveries among archaeologist Louis Leakey's famous field research in Africa were unique hominid fossils that were found together with simple hand tools made of basalt, quartzite, and bone, dating to about two million years ago.[34] The evidence strongly indicates that the hominids that later became *Homo sapiens* were upright toolmakers—*Homo habilis* ("handy man"). The findings from Leakey's research—together with the fact that all the great apes originated in Africa and most still live there—provided strong empirical support for Charles Darwin's belief that the roots of humanity can be traced to that continent—not to Asia, as had been imagined by some early-twentieth-century evolutionists. Leakey's belief that toolmaking separates humans from other species, however, has not held up well to the evidence.

It is true that advanced toolmaking marks a crucial difference between humans and other animals and represents a key stage in the evolution of our species. But toolmaking talent is by no means unique to humans, not even to vertebrates. Antlion larvae (sometimes known as "doodlebugs" because of the winding trails they make in dirt) dig tunnels and wait for prey to fall in, blocking the victims' escape route with debris. Wormlions, another larvae species, capture ants with a similar strategy. Hermit crabs occupy the shells of dead snails for protection. Marine crabs fight off predators by grabbing sea anemones and using them as shields and weapons. Some ants use leaves, pieces of wood, and small, dried clumps of mud to carry food, maximizing tenfold the amount they are able to transport with their bodies alone.[35] A veined octopus species residing off the shores of Indonesia collects and prepares coconut shells for shelter, even storing them for later use, adding another dimension to their tool use.[36] Crows and herring gulls break open shellfish by dropping them from great heights. East African vultures pick up stones and hammer the shells of large eggs until they break.[37] Some birds grasp twigs with their beaks to probe tree branches for insects. Many species of birds crack open nuts and seeds by slamming them against hard surfaces.

Not by coincidence, our closest genetic relatives exhibit the greatest ability to use tools and to communicate, skills that are closely connected. Bonobos, chimpanzees, and orangutans use twigs and blades of grass to dig insects out of wood and dirt. Orangutans employ a different kind of stick to remove seeds from fruit. Chimps and some monkeys select particular rocks and stones for the task of cracking open nuts, a skill they acquired thousands of years ago with no human influence.[38] Some male chimpanzees systematically tear leaves bit by bit between their open legs when they have an erection in order to make a rustling sound that draws a female's attention to their ready-for-love condition.[39] Chimps can be taught to pound out simple commands on a computer keyboard or, like dogs, can learn how to respond to verbal commands. Chimps can convey concepts with sign language,[40] turning their palms upward to beg humans for food, for example. They signal other chimpanzees for food, sex, grooming, or help in a fight.[41] Chimpanzees display empathy in the form of caring, mourning, and comforting, and have good short-term memory.[42]

From simple stone utensils and weapons to sophisticated computers and software, tools embody tremendous significance as instruments that represent survival potential. Being functional is highly meaningful in evolutionary terms, so the tools themselves become powerful communicative signs. The history of tools and toolmaking in many ways constitutes the primary trajectory of human culture. School textbooks, museum exhibits, and documentary videos celebrate all kinds of tools as the accomplishments of technological evolution. Most important, the evolutionary payoff for successful toolmakers, tool owners, and sophisticated tool users has always been crucial. They become valued members of communities that depend on their functional expertise and their creative talents, thereby earning social status and excellent standing as candidates for sexual reproduction.

BRAIN AND BODY CHANGES

Fueled by the cognitive demands of toolmaking, material invention in general, the need to organize socially, and a change in diet that began to include protein-rich meat, the hominid brain started to grow significantly in size sometime between two and four million years ago. It is difficult to pin down the period when our ancestors first began to use tools. Some argue that Lucy and her *Australopithecus afarensis* kin may have used stone tools to cut up animal carcasses some three and a half million years ago, though it's not clear if the sharp stones were made or found.[43] The oldest confirmed fabricated stone tools—pieces of rock with sharp edges made by cracking stones together to make flakes—date back to only about two and half million years ago.[44] Because the increase in cognitive capacity corresponds positively with tool use, it also becomes hard to determine when our brains began to develop significantly. The time when toolmaking and the major increase in the cognitive capacity of our ancestors began may never be known.

More recent history about human cognition is less opaque. As naturalists Donald Johanson and Blake Edgar write, the "biggest burst in brain size increase occurred during the Middle to Upper Pleistocene period (around 500,000 to about 100,000 years ago)."[45] The enlarged brain helped enable the acquisition of language, which stimulated additional growth in brain size, leading to progressively greater language ability. Greater language facility contributed to the development of all our advanced survival skills, especially to the ability of unrelated individuals in growing populations to work together peacefully, creatively, and productively. The tremendous advantage afforded by advanced communication skill became inscribed in our DNA. Researchers have discovered that a gene—*FOXP2*—influences the capacity to learn and use language by orchestrating the functionality of a network of genes.[46] This combination of genes, even undiscovered ones, affects the basal ganglia, the part of the brain involved in speech and language.

Inside the brain, billions of neurons constantly respond to physiological and environmental stimuli as communicating nerve cells. Much of the neurological firing takes place in the cortex—Broca's area and Wernicke's area, specifically—where the brain effectuates the production and comprehension of language and speech. Charles Darwin began to figure this out with only a scant geological and fossil record and no knowledge of DNA to guide his thinking. Darwin insisted that improved cognitive ability—including the "powers of observation, memory, curiosity, imagination"—drove all aspects of hominid development, "even at a very remote period . . . enabling him to invent and use language, to make weapons, tools, traps . . . whereby with the aid of his social habits, he long ago became the most dominant of all living creatures."[47]

The erect physical body also had to continue changing so that speech could evolve. The larynx gradually descended, opening up the vocal tract so that individuals could utter a wider variety of sounds. The tongue could now move more freely. Teeth got smaller, creating more space for articulation. Oral cavities became bigger and more resonant, enabling the production of nuanced and rapid vocalizations. These physical adaptations facilitated much greater variety in sound production, which, together with cognitive development, led to increased semantic complexity. Examination of Lucy and her "baby," Selam, reveal that the hyoid bone located at the base of the tongue—crucial for the production of speech—was intact in the bodies of our ancestors more than three million years ago. Hearing and cognitive decoding skill improved in tandem. It is not only the ability to produce complex speech but also being able to recognize and interpret utterances that makes humans so different from all other animals today.[48]

It took us a long time to get here. Primates, including humans, have been evolving for more than thirty million years—the vast majority of that time in Africa. While humans began to slowly diverge from apes fourteen to six million years ago, the most dramatic physical changes in the hominid lineage took place five to three million years ago. The *Homo sapiens* line of descent separated from the Neanderthal line in Africa between 700,000 and 300,000 years ago, meaning that Neanderthals are our cousins, not our ancestors. Anatomically modern humans evolved in Africa about 200,000 years ago, and the roots of modern behavior date back 100,000 years.

Homo sapiens began to leave Africa between 65,000 and 50,000 years ago. In *The Complete World of Human Evolution*, Chris Stringer and Peter Andrews explain that it was during this time that the species began to "colonize every inhabitable region of the planet and eventually even travel beyond it."[49] The initial move out of Africa "should not be considered as a purposeful pioneering colonization of new territories," the authors warn, "but rather as a gradual extension of their foraging ranges, tracking plant or animal resources into previously uninhabited territories."[50] The ancestral group from which the first emigrants derived was probably just a single band of hunter-gatherers numbering about 150 individuals.[51] They likely arrived first in Eurasia about 65,000 years ago, and some of their descendents continued on to Europe sometime between 45,000 and 35,000 years ago.[52] Early European modern humans lived alongside the Neanderthal until our cousins went extinct between 35,000 and 30,000 years ago.[53]

The formation of ancestral human populations, the exit of subsets of that population from Africa, and the extensive development of language seem to have all taken place around the same time.[54] Language and the social cohesion that communication ability facilitates gave the first humans to leave Africa a decisive advantage over other hominid species that emigrated thousands of years earlier. Over time, the accumulation of communication acts strengthened social ties that helped ensure the survival of individuals enmeshed in the nomadic groups. Besides articulate speech, the migrating human populations likely had developed spiritual beliefs, ways to defend themselves, and a basic sense of social reciprocity even before they left Africa.[55] Communication skills made it possible for them to coordinate hunting, to care for offspring, and to create viable and increasingly complex communities. All contemporary languages descended from this primordial tongue or tongues as people voyaged in all directions across the face of the earth. Some form of vocal communication developed everywhere. As neuroscientist Michael Gazzaniga points out, in the entire history of human evolution "there are no reports of a newly-discovered band of hunter-gatherers who lacked language, only to suddenly learn it from their more technically-advanced neighbors."[56] Dispersed human populations not only communicate similarly, but they also act pretty much the same way in general.[57]

LANGUAGE SKILL

Convinced that an unbroken continuity stretches from nature to humans to culture, and that close links connect lower-order with higher-order animals, Charles Darwin surmised that language "owes its origin to the imitation and modification of various natural sounds, the voices of other animals, and man's own instinctive cries, aided by signs and gestures."[58] All social organisms have communication systems. Sentient beings exhibit the greatest variety. They "speak" with hisses, barks, chirps, whistles, screams, whinnies, bellows, and hundreds of other sounds and nonverbal expressions. Communication is by no means uniquely human, but the capacity to intentionally express an infinity of meanings by using a small number of distinct sounds, arranged creatively and precisely to represent original, complex, and abstract ideas, is peculiar to our species.

Does all this evidence suggest that language skill is hardwired into the brain? Linguist Noam Chomsky wrote nearly fifty years ago that every human being is born with an innate sense of the rules for linguistic expression—what he called "universal grammar."[59] People don't learn how to speak; they know how to do it from birth. An internal language template in the brain initiates and guides communicative expression so that the ability to speak a language results naturally from biological maturation. The specific rules and vocabulary of language derive from contextualized experience, but, according to Chomsky, the deep structures of language are built into our biology as a specialized brain module.

If a language module exists in our brain, how did it get there? Cognitive scientist Steven Pinker and others believe that universal communication ability resulted from selection processes that took place in the early stages of human evolution—long before *Homo sapiens* began their journey north. Language development, social development, and cultural development evolved interactively, each contributing to the growth of the others. For Chomsky, the key to understanding language development lies in neural hardwiring given by nature; when we begin to speak as young children, we activate built-in circuits. For Pinker, those very circuits are produced by more than two million years of evolutionary history and continue to be influenced by selective processes adapting to the particular environmental and cultural conditions that surround us today.[60]

Deep structural characteristics underlie all languages, but each has its own grammar, which often differs considerably from the structure of other grammars.[61] Cultural commonalities in the way languages develop stem from shared motives to communicate and from similar patterns of cognitive and social development among human groups. But from that point forward, variation takes over. Even within distinct speech communities, linguistic practices give way to the expressive idiosyncrasies of individual persons, as Canadian anthropologist Grant McCracken describes:

> Each speaker of a language is both constrained and empowered by the code that informs his language use. He or she has no choice but to accept the way in which distinctive features have been defined and combined to form phonemes. He or she has no choice but to accept the way in which the phonemes have been defined and combined to form morphemes. The creation of sentences out of mor-

> phemes is also constrained, but here the speaker enjoys a limited discretionary power and combinatorial freedom. This discretionary power increases when the speaker combines sentences into utterances. By this stage the action of compulsory rules of combination has ceased altogether.[62]

Evolution has added layer upon layer of discrete and combinatory forms of digital linguistic expression to our primordial repertoire of analogical signals, allowing us to represent very complex mental states. Digital messages are composed of a system of oppositions. Contrast identifies each element. Speakers must utter the correct sound or display the right image to deliver the message clearly. To use a simple example, imagine you are on a baseball field talking to the other players on your team. You ask, "Is this your bat?" Your teammate replies, "No, it's not mine." Had you inquired, "Is this your cat?" the message would have been completely senseless. Both the speaker and the receiver must have it in mind that both "bat" and "cat" are possible utterances in any communicative exchange, otherwise no choice would be made, or the response would be random. Yet in a microsecond, you chose to vocalize a *b*, not a *c*, to begin the critical word, because *bat*, not *cat*, was the intended term. It was instantly recognized by your interlocutor. We make subtle and discrete decisions like this with every sound we utter, usually without thinking about it. So, as language became more intricate, systematic, and exact, cultural information could be passed from one person to another with increasing accuracy and influence.

Spoken and written language is but one type of digital code. Consider another familiar code system: traffic lights. Each color that appears on a traffic semaphore is placed in opposition to the other colors, forming a simple communication system. Green (+) is the opposite of red (-) with yellow signaling a condition between the extremes (+/-). Based on a closed system of oppositions, three meanings can be constructed: go (+), stop (-), and watch out (+/-). Because the binary system denotes discrete shared meanings and the semaphores illuminate in an order that creates and then constitutes shared knowledge, a color-blind person can correctly read any traffic light without actually discerning the colors. Digital message systems are flexible in this way; they need not be restricted to what they conventionally represent. But once established, digital systems become naturalized and are difficult to change. Breaching the system can be counterproductive, even dangerous. During the Cultural Revolution of the mid 1960s, the Chinese national government considered inverting the Western system of traffic signaling by placing red at the top and green at the bottom. Over the long term, the changeover would have become viable, but adjusting to the new system would have caused numerous accidents and may have cost hundreds of Chinese people their lives!

We compose messages from an endless assortment of linguistic and nonlinguistic elements that we creatively combine to explain complex thoughts and feelings and to draw fine shades of meaning. We describe the intricacies of situations, discuss possible solutions to difficult problems, and pontificate about things we can only imagine. We recursively embed multiple subjects and temporal increments within a single utterance. We use nonverbal signaling to reinforce what we say with language and send messages about our deepest needs, wants, and feelings. We gesture even when no one can see us—while talking on the telephone, for example. Animals also use sounds, gestures, and facial expressions to communicate, but animal communication doesn't manifest anything close to the levels of reflexivity, intentionality, combinatorial complexity, and subtlety that humans employ.

Instincts that have evolved over countless generations—like knowing the right times and paths for migrations—prompt animals to take enormous risks. They remember past actions like where they buried food. Members of species who share food remember and take care of other individuals in their communities who have a track record of sharing. Sexual favors granted by female primates reward past contributions by specific males to the female's well-being in the form of food and grooming. While these behaviors have also evolved as instincts, the particular recipients of rewards are those individuals who have been identified and remembered as deserving of reciprocity. Animals make many decisions about what to do in the here and now according to the dictates of short-term practical memory.

Animals have a sense of time inscribed in their DNA. Instincts deriving from and leading to repeated behavior contribute to an individual's survival; practical memory prompts specific behaviors that respond well to the demands of a particular situation. But many animals also forecast future actions—the rapid changes of direction of prey under the pressure of pursuit, for instance. Storage and retrieval of food is a common cross-temporal behavior. Chimpanzees can make plans that take into account the likely future actions of others, even of humans. At a zoo in Sweden, for example, chimps stockpile stones and rocks with the intention of throwing them at visitors.[63]

Although the forces that trigger their communicative activity draw from the past and consider the future, the messages that animals send refer only to what's happening in the here and now.[64] Humans, on the other hand, routinely communicate about objects and events in the past, present, and future. This ability, known as displacement, functions in concert with an almost infinite set of signals and signal combinations, to give humans a considerable evolutionary advantage over all other species.[65]

The depth and flexibility of human communication allows us to focus on, to contemplate, and to criticize everything around us. Among other motivations, humans communicate to create complex social systems, reflect critically on our actions, and take the perspective of others into account—even the views of individuals we don't know or with whom we disagree. We solve problems by conversing and negotiating. Humans develop theory of mind—the capacity to put one's own situation into perspective and infer what others know or don't know and what they might or might not do—early in life. This ability exists only in a very basic form in other primates, if at all. Animals don't form extended symbolic sequences. No other animal communicates about how it communicates. Multiple levels of social sensitivity, the capacity for introspection, and the ability to negotiate a course of action in response to feedback—flexibility—endow human beings with enormous power to positively influence their cultural development.

FROM PREY TO PREDATOR

Many people think evolution proceeds sequentially and progressively. Scientific sketches and popular graphics show quadrupedal hominids morphing step-by-step into bipedal modern humans. Comical variations take the drawings further—we evolve into skateboarders, computer geeks, golfers, and iPod® users. A rosy optimism always pervades the image, contributing to the reassuring perception that in the Evolutionary War, humans come out on top. But in the colossal struggle to survive, have humans always had their way? Does scientific evidence support the idea that we were born natural predators?

That was the prevailing view of physical anthropology through the early part of the twentieth century, and it was backed by mainstream religious ideology.[66] Sherwood Washburn, a leading American scholar of the period, was so confident of humans' natural predation, he repeated the same strong claim in multiple publications: "Unless training has hidden the natural drives, men enjoy the chase and the kill. . . . The victims may be either animal or human."[67] Australian anthropologist Raymond Dart agreed that the progenitors of modern humans slew and ate not only other animals but their brethren as well. He based his reasoning on the presence of mutilated antelope carcasses and human skulls found in African caves, doubting that the bones were the remains of local predators' victims.[68]

Scientists in the early twentieth century were producing exciting field research supporting the idea that humans had evolved over millions of years. Yet religious faith was strong for many of the scientists and for the vast majority of people in their societies. How, then, could the mounting empirical evidence pointing to evolution be reconciled with firmly entrenched religious belief? The title of Louis Leakey's important book *Adam's Ancestors: The Evolution of Man and His Culture* (1934) reveals the hedge that was often made. Born to Christian missionaries in Kenya and a devout believer himself, Leakey made the discovery of hominid fossils in Africa that helped piece together the true story of how and where human life began. But his description of our evolutionary past incorrectly gave *Homo sapiens* an exalted place in the natural world. Leakey posited that humans branched off from all the living apes more than thirty million years ago. He placed all other hominid fossils (Java Man, Peking Man, the Heidelberg Jaw, Rhodesian Man, and the Neanderthals) far away from the line of human ancestry.[69]

Raymond Dart used religious rhetoric to express his claim that humans' natural predation helped them dominate in the struggle to survive. He began a widely influential article by quoting seventeenth-century Calvinist preacher Richard Baxter from his book *Christian Ethics* by saying, "Of all beasts the man-beast is the worst . . . to others and to himself the cruelest foe."[70] From this view, humans should be considered as flesh eaters right from their primate origins and as the fiercest of all animals. But, according to Dart, humans had to be tamed. Fortunately, a solution was available: submit to the Lord and let him harness humans' destructive instincts.

Washburn, Leakey, Dart, and other scientists and intellectuals must have known that the creationist claims of the Old Testament, accepted without question by all three major monotheistic faiths, were not literally true. Still, some accommodation had to be made for religious belief. The Book of Genesis and its description of the Fall of Man provided the justifying narrative: man, the lamb, the lion, and every other creature lived together peacefully in the Garden of Eden until Adam disobeyed the holy dietary prescription and shattered the divine order of nature. The Fall ensued, and man began to suffer the sin of predation.[71] Jesus gave up his life for our sins centuries later, condemning man as innately immoral. For Washburn, Leakey, and other Western scientists, this alarming morality tale originated in Christian folklore, but a similar story is found among many other belief systems, including Greek mythology (the resurrection of Dionysus) and touches on many themes in pagan folklore.[72]

But what if those early mortal beings were at first not predators but prey? Unfortunately for our primitive ancestors, that appears to be the case. Teeth marks on human bones found in the late twentieth century in South Africa, Ghana, China, and elsewhere indicate that early humans were attacked and devoured by fierce predators—saber-toothed tigers, leopards, hyenas, even huge birds.[73] Why would we argue with this evidence? Our closest primate relatives, chimpanzees, are still routinely preyed upon by hyenas, wild dogs, lions, leopards, and other carnivores in the wild. Humans are medium-sized, slow-footed mammals that are more physically vulnerable to predation than they are naturally capable predators. "All apes and humans are primarily fruit eaters," Chris Stringer and Peter Andrews point out. "Chimpanzees and humans also eat animal flesh, but neither has any particular biological adaptations for catching and eating other animals."[74] Chimpanzees still protect themselves by building nests high in trees and not foraging beyond safe limits. Thus the evidence has become increasingly clear: humans exist on the primate continuum, not above it. The first humans were the likely prey of cunning and powerful predators, not dominant predators themselves.[75]

Today, of course, superior cognitive development and weapon-making ability allow humans to bring down even the largest and fiercest beast. Instinctually, though, predators still threaten us. Nature lovers, farmers, loggers, poachers, and rural residents all over the world have to watch out for large carnivores when venturing into the wild. The most dangerous region in the world may be the densely forested delta of the Genghis and Brahmaputra Rivers between India and Bangladesh. Human-eating tigers killed more than six hundred people in that area from 1975 to 1985, and predation by vicious felines in similarly dangerous areas of Asia and Indonesia continues today.[76]

Big felines—tigers, lions, leopards, jaguars, and cougars—aren't the only animals that prey on humans today. When under severe environmental pressure or when they sense a rare opportunity, alligators, crocodiles, sharks, bears, wolves, and pythons will attack and consume humans, especially small children. In late 2009, young Canadian folk singer Taylor Mitchell was viciously bitten to death by two coyotes as she hiked alone in a national park in Nova Scotia, prompting international news coverage. A year later, thirty-two-year-old special education teacher Candice Berner was attacked and killed by two or more wolves as she jogged along a deserted road in southern Alaska, listening to her iPod. Her partially devoured body surrounded by wolf prints was discovered by snowmobilers soon after the attack took place. In search of food, a polar bear attacked a group of young campers in the Arctic region of Norway in 2011. Wildlife attacks by bears, wolves, and moose are relatively common in rural Alaska. People everywhere become morbidly fascinated by such stories, and few question the truth of the accounts.

In order to avoid death when confronted by bigger and more powerful beasts, humans have to win a contest of wills by communicating dominance. For example, wildlife authorities tell hikers and campers that upon encountering a mountain lion they should establish and maintain constant eye contact with the cat. The person should stand up tall, even hold a jacket above the head so as to appear very big and make loud sounds, and never to crouch, hide, or run away as a fearful creature might do.

Authorities in the dangerous Indian delta region greatly reduced the number of tiger attacks by employing another communication strategy. The officials issue plastic masks to people who travel the area by boat. The masks are worn on the back of the head and feature large bulging eyes that signal awareness and ferocity to predators on the prowl. Because big cats want to attack by stealth, they become discouraged and move on to better prey. The mask strategy appropriates recurrent evolutionary outcomes that take place in the wild. They serve the same defensive purpose as conspicuous eyespots on the wings of butterflies for keeping birds away. Protruding, brightly colored eyes on the red-eyed tree frog startle snakes and birds. Big eyespots near the functioning eyes of oyanirami fish turn away underwater predators.

Predatory roles seem very well proscribed. The wildebeest, an herbivorous African antelope, will not react to a vicious attack by a pack of strong-jawed hyenas by trying to devour the smaller predators.[77] A zebra does not fight back against a pouncing lion with any intention of eating the attacker, even though the zebra may weigh twice as much as the lion. Cows never reverse the role of their common predators—gray wolves and coyotes in the United States; cheetahs and panthers in Africa. A host of one-way predatory relationships has become very familiar through observation and folklore—the fox and rabbit, the cat and mouse, or the bear and fish, for example.

Long-standing predator-prey patterns run deep in our evolutionary history. The fear of predators never fades away completely, even for people who live in great cities.

The vast majority of prey is no match for the predators that besiege them. How was it possible, then, that early humans evolved from vulnerable prey to dominant predators—from the hunted to the hunter? Some scientists believe humans became successful predators after their cognitive capacity expanded and they acquired new survival skills and technologies—especially the mastery of fire and the making of simple tools and weapons.[78] Others point to the fact that catastrophic climate change and migrations forced humans to innovate in every respect just to survive.[79] These developments forced evolutionary adaptations, but they alone could not have altered our position in the food chain. Something else must have occurred to account for the extraordinary trajectory of human evolution.

Humans became the dominant species in the animal kingdom mainly because they developed a progressively adaptive system for exchanging thoughts and instructions. Primitive cultures collaborated to create weapons for hunting and to exchange information about the prey they stalked, and experimented by hunting in teams where individuals assumed diverse responsibilities. The survival value of these interrelated elements—weapon development, transmission of information, and social coordination—has never diminished. Even the most recent innovations in communications technology have their roots and applications in the quest for military domination. This was the case for the development of the radio during World War I; for the television in World War II; and for information technologies during the Cold War, the Vietnam War, and the wars and recent surreptitious interventions in the Middle East. The Internet was created from investments made by the Advanced Research Projects Agency (ARPA) of the US Department of Defense in response to the launch of Russia's Sputnik satellite in 1957.

But the story of our evolutionary success does not rest solely on our predatory qualifications. Primitive cultures did not develop as completely male-dominated groups whose only incentive for cooperating and communicating was to establish and defend territory or to hunt and consume animals. The caring, nurturing role of the female has a long evolutionary history, especially among mammals. Communication skill enables the supportive behavior. Primitive cultures did not develop solely as male-dominated groups whose only incentive for cooperating and communicating was to establish and defend territory or hunt and consume animals. Humans evolved as cooperative breeders (including child raising, not just procreation) among whom extensive social relationships function to evolutionary advantage. Child rearing is undertaken not only by the birth mother but also by other family members, distant relatives, and neighbors. A profound sense of social responsibility, relationships built on trust and reciprocity, and effective interpersonal communication are needed to connect nuclear families in their various configurations with other caregivers. The instinct for cooperative breeding was established in response to one of the most compelling forms of human communication—the sounds and the facial and bodily expressions made by infants. Their disarming behavior encourages people to share, to work together with strangers, and to become more attentive to the needs of others overall.[80]

Whether motivated as toolmaking, chest-thumping predators or other-directed, trusting caregivers, human development requires a degree of social cooperation that could have been facilitated only by the advent of speech and language. David Sloan Wilson, who has made a welcome effort to bring the study of human evolution into mainstream college teaching and popular consciousness in America, writes of the "three C's of human evolution: cognition, culture, and cooperation."[81] But for all his positive contributions, Wilson came up one "C" short and got the order partly wrong. No communication? Then there can be no cooperation. No cooperation? Then there can be no culture.

The transformative role of communication underscores a point Charles Darwin described when he formulated his theory of change: evolution proceeds as a relentless zigzag, an irregular weaving and unweaving of endless possibilities, a parade of hopscotch jumps that leads to places biology alone cannot determine. Evolution is a balancing act on a high tightrope with no safety net. While most biological mutations go nowhere, some alterations—like the first gestures and sounds expressed by our hominid ancestors that would evolve into language and culture—have proved to be the adaptations that shaped our history more than any other single influence.

ARTFUL EXPRESSION

The enormous symbolic power wielded by today's global pop culture stars—even the most indelicate among them—wouldn't have surprised Charles Darwin. He observed that singing, music making, dancing, poetry, painting, tattooing, body painting, and body piercing pervaded diverse populations long before modern media and the culture industries made them trendy. Believing that the deep structure of artful expression derives from tribal roots, Darwin was fascinated by indigenous peoples' "universal habits of dancing, masquerading, and making rude pictures."[82] Tribal cultures in Darwin's day disfigured, scarred, punctured, pierced, stretched, and painted their skin as a medium for bodily expression—customs that persist today in myriad forms and certainly not just among indigenous groups.

Darwin reasoned that artistic expression emerged among human groups in much the same way emotional displays, gestures, and spoken language evolved in cultures around the globe—and that all these developments are interrelated. Later research proved him right. The production of early symbolic forms took place independently and in some cases concurrently throughout the world. For instance, aboriginal art in Australia and North America—completely unaffected by each other—both date from the same period.

Love of novelty, Darwin thought, lies at the heart of all forms of expression and contributes greatly to universal cultural development. Even animals "appreciate slight changes in colors, form, or sound," he said.[83] But there must be a more comprehensive explanation for why art, music, dance, fashion, and symbolic language grew to be such important cultural features in all human societies.[84] What motives, for example, could possibly keep the various culture and fashion industries—which depend on ceaseless novelty and change to survive—so vibrant, profitable, and enduring? What is the adaptive value of artful expression?

Reproductive potential is one prime motivator. People desire mates who are healthy and beautiful. The expressive body demonstrates crucial dimensions of individual fitness—originality, creativity, aesthetic qualities, physical skills, and stamina among them. In modern Western societies, the ways men and women present themselves to others send vital signals to potential sex partners but also to employers, colleagues, family members, neighbors, and clients who—possibly unconsciously—render consequential judgments of fitness and sexual attractiveness. As anthropologist Nina Jablonski describes it, "Self-decorating . . . is all about attracting a mate and getting a chance to reproduce."[85] For these reasons, she points out, women everywhere use makeup like eyeliner to make their eyes appear larger—a trait considered to be an appealing physical feature almost everywhere (think eye surgery among Asian women, for example). Men alter their skin with tattoos and piercings and make fearsome facial expressions to enhance their desirability as sufficiently masculine mates.[86] Even the most primitive forms of artful expression reflect one of communication's primary roles: to attract and hold the attention of others.

Artful expression serves other purposes, too. Apart from their skin, the first surfaces painted by humans were the walls of caves, an activity that began at least 75,000 to 80,000 years ago during the Middle Stone Age. On the walls of South Africa's Blombos Cave, researchers discovered markings made in red ochre—iron oxide that is mixed with clay or sand to make bright shades of red, orange, or yellow. Also found was an ochre workshop, containing grinding stones for making the pigment, abalone shells for containing the paint, and bone spatulas for scooping the liquid onto surfaces. The groups who fashioned these tools demonstrated the ability to find, combine, and store materials for later use—a key development in human cognition.[87]

Sticks of ochre were the first known natural tools employed for the purpose of symbolic representation. The same kinds of materials—ground minerals, plant substances, blood, charcoal, and urine—were used to make the dark red colors indigenous tribes of America's Southwest employed thousands of years later to paint pictographs on the walls of rock dwellings and other flat surfaces. Cave paintings represent an extraordinary stage in the evolution of human communication. For the first time, thoughts and information were displayed physically outside the human brain and body, changing cultural perceptions of time and space along the way.

The South African caves contained another key artifact: strings of manually perforated shells that experts believe served as ornamental beads. The shells were strung together so they could be used as necklaces or bracelets. Similar decorative shells of an even earlier vintage have been discovered in Israel and Algeria, suggesting that the first forms of human expression external to the human body may have originated as long ago as one hundred thousand years or more.[88] Human communication and social interaction generally were becoming more complex. Toolmaking and jewelry making both require the transmission of knowledge from one individual to another. Redundant messaging was used to make sure the learner understood the process.

These ancient cave markings and jewelry lend further credence to the argument that early humans were not only making tools and speaking a protolanguage before leaving Africa and the Levant; they was also creating simple forms of art. The Blombos Cave artifacts and the subsequent discovery of ornamental shells in Israel and Algeria call into question the idea that a sudden and unprecedented creative explosion in symbolic production took place after the first humans left Africa. A more gradual development of artful expression more likely transpired.

Yet the abundance of material evidence dating from the Upper Paleolithic period (45,000 to 10,000 years ago) reveals that a notable increase in cultural production did in fact occur at a later time. Early modern humans were changing from nomadic hunter-gatherers to permanent farmers—from finding food sources to creating them. Toward the end of this period, the crops developed by these early humans—corn, wheat, and rice—and the animals they domesticated as food sources led to changes in their diets, which in turn led to dramatic effects on their bodies.[89] Settling down, growing grain, and domesticating animals also transformed the meaning of time and allowed for more innovation and expression. Tools became more complex. Hunting, trapping, and fishing techniques grew more sophisticated. Fire technology and cooking methods improved. New domestic instruments and practices, including the making of clothing and the fashioning of more elaborate living areas, appeared. Populations increased in size and longevity. Vocational specialization, more specific gender roles, and social hierarchies emerged. Intricate trade alliances and other social networks surfaced—developments that would lead to the origin of writing many years later.

Representational art—pictographic sketches and stone etchings known as petroglyphs—proliferated at pace with the social and cultural changes. The most impressive and well-preserved evidence from this era can be found in the cave art of southwestern France, eastern

Spain, southwestern Germany, and northern Italy. The most common subjects were horses, bison, and deer, but exotic creatures like mammoths, rhinoceroses, and ibex were also represented. Portraying images of these beasts suggests that early humans may have thought they had mastered nature.[90] The first schematic representations of the human physique were also drawn on cave walls during this period—often in hybrid form, in which the male body is fused with the head of a lion, for example. These depictions are the work of the first culturally modern human population.

Some of the southern European caves also yielded small figurines of stone and bone, ivory sculptures, clay statuettes, jewelry, and primitive musical instruments. Beads and pendants made from shells, stone, amber, and mammoth ivory were fashioned by early Western populations into buttons, necklaces, bracelets, and decorative clothing.[91] The oldest of the more substantial artifacts—a provocative ivory carving some 35,000 years old—was recently unearthed in Germany.[92] This latest nude "Venus" sculpture, similar to female figurines from later periods, was carved to dramatically emphasize the model's sexuality—huge protruding breasts, big buttocks, and an extremely enlarged vulva. We get a strong message from this and other Stone Age art about the signifying power of sex in human culture and the first symbolic objectification of the female form.

Early modern humans used natural materials to make music, too—and some of their materials and methods are still employed in modern times. Children today make musical sounds by blowing through hollow sticks or leaves they hold between their palms, by beating surfaces with small branches, or by rattling stones together. The first flutes were made by poking holes in animal and bird bones. More durable wind instruments were fashioned from bone and ivory. The earliest evidence of human music making was uncovered along with the pictographs, jewelry, and sculpture in the caves of Europe and dates to the same time period—roughly 35,000 years ago.[93]

Cultural production since then certainly has been robust. But what are the evolutionary functions of art? Is art directly adaptive, providing tangible benefits that help individuals to survive and reproduce? We know, for instance, that creative, expressive individuals have more success than others finding mates and passing on their genes to future generations.[94] It's also clear that art provides ways for groups to express their histories, priorities, and spiritual musings. But art can also be considered a by-product of more central evolutionary processes, a side consequence that didn't become useful until later.[95] It can be argued, for example, that as an accidental consequence of vocalization early humans also learned to sing simple melodies, which became a useful adaptation for lulling babies to sleep or for reinforcing communal ties in religious singing.

It could also be the case that there is so much beauty, joy, pleasure, and wonder in art—all its forms—that we simply cannot account for its universal importance by thinking of it only as either a functional adaptation or as a by-product of selection pressure. This aesthetic-based view is what art philosopher Denis Dutton means by the "art instinct." Prehistoric women and men may have been charmed by beauty in and of itself and were imbued with a natural desire to express themselves artistically. According to Dutton, "The aesthetic arises spontaneously as a source of pleasure in cultures across the globe. . . . How strange to argue for a Darwinian genesis of the arts of man, which so often tend toward lavish excess, costly far beyond any obvious adaptive benefits for our survival."[96]

Art communicates. It is appreciated everywhere as aesthetically pleasing and entertaining. But art forms have always performed as social media as well. From the grunts and groans uttered by our hominid ancestors to the spoken languages, music, cave art, jewelry, writing, and right up to the full complement of modern digital media, the need to survive and reproduce has inspired people to create and communicate. *Homo sapiens* art became portable and eventually went global with positive effect. Indeed, advanced expressive skills—including literacy with a diverse range of cultural media—have given *Homo sapiens* an overwhelming evolutionary advantage. The advantage is but a matter of degree, however. Our closest hominid relatives and past competitors, *Homo neanderthalensis*, also had big brains and social abilities. They made bone and stone tools and probably had a simple protolanguage. They may have sung to heal and calm their offspring.[97] But for whatever reason, the Neanderthals never developed an ability to express themselves more fully. They left no art or symbolic artifacts other than simple tools. For these reasons, it is often argued that limited communication skills contributed significantly to the Neanderthals' demise thirty thousand years ago.

UNVEILING NATURE

In early June 1834, Captain Robert FitzRoy carefully navigated the HMS *Beagle* with Charles Darwin on board through the Strait of Magellan, skirting the northern edge of Tierra del Fuego, off the southernmost tip of South America. The captain was getting ready to turn the ship northward along the Chilean coast toward the Galápagos Islands—where Darwin would collect biological specimens that forever changed our view of the world.

As the captain prepared to make the turn, Darwin was disappointed that Mount Sarmiento, the spectacular Andean peak situated at the point where the Magellan Strait bends up the coast, had been shrouded in fog for days, making it impossible to see. But the weather changed on June 9, slowly revealing an unforgettable sight. The young Darwin recorded the event in his diary: "In the morning, we were delighted by seeing the veil of mist gradually rise from Sarmiento, and display it to our view."[98] As the clouds dispersed, Darwin and his shipmates cast their eyes on a breathtaking rocky prominence of nearly seven thousand feet that Darwin described as "clothed by dusky woods, and above this a field of snow extends to the summit. These vast piles of snow, which never melt, and seem destined to last as long as the world holds together, present a noble and even sublime spectacle."[99] The veil had been lifted from the mountaintop; the awesome beauty of the cordillera's southernmost crest revealed.

Beauty in all its natural forms made up an integral part of Darwin's personal view of life and, ultimately, his magnificent theory of evolution. More than twenty years after the *Beagle* returned to London, he phrased the conclusion of *On the Origin of Species* with a famous thought: "There is grandeur in this view of life . . . from so simple a beginning endless forms most beautiful and most wonderful have been, and are being, evolved."[100] Evolution has no intention or purpose. Yet Charles Darwin's appreciation for "endless forms most beautiful and wonderful" reveals a deep and hopeful spirit that embraces the idea of "progress toward perfection" that he believed evolution offers humankind.[101]

CHAPTER 3

COMMUNICATING SEX

In *The Social Network*, the iconic 2010 film directed by David Fincher that examines Communication Age entrepreneurship, Mark Zuckerberg, eventual cofounder of Facebook®, has a revelation. As he's working away at a computer science lab at Harvard University, his friend Dustin Moskovitz approaches him with a question: "There's a girl in the art history class that you take. Her name is Stephanie Attis. Do you happen to know if she has a boyfriend?"

The perpetually wired-in Zuckerberg looks annoyed by the interruption.

Moskovitz persists: "I mean, have you ever seen her with anyone . . . do you happen to know if she's looking to go out with anyone?"

Zuckerberg stops typing and stares off into space for a moment. He suddenly jumps up, grabs his backpack, and flies out of the building. Barely able to keep his balance on the winter ice, the young student races across campus and up two flights of stairs to his dormitory room. His roommate, Eduardo Saverin, says something to him as he bursts into the room. Zuckerberg ignores him and opens his laptop.

"I have to add a feature," Zuckerberg says. "Relationship Status . . . Interested In"

He explains to a quizzical Saverin: "These two things are what drive life at college. Are you having sex or aren't you? It's why people take certain classes, sit where they sit, go where they go, do what they do, and . . . that's what the Facebook is gonna be about. People are gonna log on because after all the cake and watermelon there's a chance they're gonna . . ."

"Meet a girl," Saverin finishes the thought.

"Get laid. Yes," Zuckerberg says. "It's ready."

"It's ready?" Saverin asks.

"Yeah, it's ready. That was it."

Facebook went live in 2004. Today, Facebook has nearly a billion active users, more than half of whom log on every day, in nearly two hundred countries. Users boast an average of 130 friends and share 30 billion pieces of content (web links, news stories, blog posts, notes, photo albums, and so on) each month.[1] Lots of what they post and chat about has little or nothing to do with sex. But who can deny Zuckerman's claim that sex is what drives social networking—online or off? And who would argue that entrepreneurs like Zuckerman aren't themselves motivated for the same reason? As Zadie Smith comments in a review of the film, "Don't we all know why nerds do what they do? To get money, which leads to popularity, which leads to girls."[2]

Less constrained than ever by geography and local culture, today's digital communications environment offers near limitless opportunities for people to exchange messages. People display social and sexual attractiveness by posting cleverly on social networking sites. They text, instant message, and e-mail each other. Some invent multiple personalities in order to create and maintain numerous relationships. The motives of people who join social networking sites are predictable: males are more likely than females to use the sites to meet females and to flirt; females typically use the sites to maintain preexisting friendships.[3] By creating the "relationship status" feature on Facebook, Mark Zuckerberg took advantage of a fundamental evolutionary principle: above all else, sexual desire drives social interaction. Today's Internet traffic confirms that evolutionary truth.

THE NATURE OF CHANGE

Charles Darwin understood the power of sex early in his work, but he chose to focus first on the grand effect of all evolutionary processes: change. From the beginning, flickers of the general insight that gradual evolution is an essential quality of nature became apparent to Darwin. But the explanation of slow and cumulative biological change that he would develop into an extensive theoretical treatise many years later was not original to Darwin, nor did he ever claim it was. Long before his voyage on the *Beagle*, scores of philosophers, physicists, and cosmologists had put forward the idea that nature had at least somewhat evolved. Even Darwin's paternal grandfather, Erasmus, himself a well-known intellectual, wrote about how species compete and change.[4] The predominant notion that our world and the universe have simply always been there in the forms they exist in today made little sense to many learned persons before Darwin's time.

Change in the natural world is a familiar theme in ancient Greek philosophy. Comparing the passages of life to the currents of a flowing river, Heraclitus of Ephesus claimed that the physical world is not stable. It changes. Evolves. As the familiar phrase has it, one cannot step in the same river twice. Plato also recognized that the universe changes across time. For Plato, change implied negativity—if something undergoes change, it must not be perfect. Early Christian thinkers such as Augustine of Hippo believed that change moves relentlessly toward the better—from the Fall to Redemption, from original sin to salvation. Because humans were created in God's image, and are thus endowed with free will, they can evolve toward perfection. Augustine separated nature from humanity. Nature was created in a flash as a totalizing projection of Divinity's plan for humankind.

Despite differences between the two philosophers—Plato waxing negative about the loss of essence and Augustine convinced that humanity can find perfection in the afterlife—their contentions share points in common. Time has a fixed direction: decaying, in the case of Plato, or improving, as Augustine thought. These philosophers believed that just because earthly souls don't comprehend how life processes unfold doesn't mean a predetermined course for humanity doesn't exist. Some kind of plan *must* be in place. Just look around, they would argue, the natural world operates like perfect clockwork with many complex parts all well fitted with each other—just as the Creator would have it.

The foundation for the theory of evolution was also grounded in a new understanding of historical time—specifically in the English geologist Charles Lyell's discoveries about the actual age of the earth. Lyell's research in the early nineteenth century revealed that the earth is much, much older than had been previously thought. Darwin took Lyell's *Principles of Geology* with him on the *Beagle*,[5] and what he saw on the voyage—marine fossils wedged high into the rocky edifices of the Chilean cordillera, for example, or the contrast in surface terrain across the Galápagos Islands—coincided with what he was reading. Deep time, he reasoned, makes incremental biological change possible.

It was becoming clear to the young Darwin that the relevant question was not *who* created this astonishingly adapted clockwork but *how* the clockwork could be explained without falling back on the belief that it was brought about by some kind of supernatural force. It may seem odd that modern biology's most radical insight could come from developing theories of nineteenth-century geology, but what Darwin was doing as a naturalist aboard the *Beagle* could, in many respects, be defined as geology. In a letter sent home from Rio de Janeiro in May of 1832, Darwin declares, "Geology cries the day; it is like the passion for gambling. Speculations, on first arriving at what the rocks may be."[6]

Charles Darwin explained the presence of constant biological change and its resulting variation by the principle of divergence, which he described in the first of his great books.[7] Nature is a perpetual work in progress. New species come into being as the modified descendants of other species. Some species proliferate while others go extinct. To explain how this cascading series of natural events has unfolded from antiquity to the present, Darwin advanced an alternative notion of temporal flow. Time doesn't travel through historical epochs like an arrow. Nor does it function like a mechanical device going forward or backward, up or down. Instead, time branches out like a tree. When we look back in the direction of origins, we find unity—the roots of life. When we glance forward to the future, we see endless speciation—the branches and twigs.

Although Charles Darwin himself and most Darwinian scholars single out his voyage onboard the *Beagle* as the starting point that led him to the idea of evolution by natural selection, such an earth-shaking idea could not have sprung from a singular experience even if it had taken place in a unique ecological niche—the Galápagos Islands. That would be like claiming Newtonian physics suddenly burst onto the scene after an apple fell on the Cambridge professor's head. Long before his voyage, Darwin must have had questions that religious orthodoxy couldn't answer.

Still, when boarding the *Beagle* in Davenport, England, on December 27, 1831, Charles Darwin clung to a kind of natural essentialism. Family history and latent religious belief probably explain why. After Darwin left medical school in Scotland due to lack of interest in his studies, his father proposed that he become a minister, an idea Darwin seriously considered. In his autobiography, Darwin reveals he "liked the thought of becoming a country clergyman."[8] He wrote, "I did not then in the least doubt the strict and literal truth of every word in the Bible. . . . It never struck me how illogical it was to say that I believed in what I could not understand and what is in fact unintelligible."[9] He packed a pistol and a Bible with him on the trip. Any personal faith he took with him, however, began to wane. Darwin recalls that as he set sail aboard the *Beagle*, he still accepted "the permanence of species, but as far as I can remember vague doubts occasionally flitted across my mind."[10]

However, it wasn't until after he returned to London in the autumn of 1836 and was preparing a journal of his trip for publication that Darwin began to develop the idea of common descent with modification—evolution. Until then, his thinking about organic creation remained remarkably akin to the biblical narrative. Species were created separately, each one following a separate course like parallel lines that never converge. The earth's diverse creatures developed the way God created them—one organism at a time.

NATURAL SELECTION

Darwin sketched out the initial idea of natural selection several years before he published *On the Origin of Species*. The explanation draws substantially from his experience as a breeder of domestic animals and plants. Working within relatively short time frames, stock breeders can improve animals by mating pairs that are selected for particular traits. The more advantageous coloring, size, health, and behavior of parents will show up in their offspring. The same basic process, Darwin thought, must take place in the wild but without any control over outcomes. The organism's interaction with the environment shapes the outcomes of reproduction. It's a very slow process that leads to diversification, not perfection. Natural selection proceeds less as a desperate search for the new than as an uncharted fiddling with prior solutions.[11]

The inherent variation in nature should eventually lead to the creation of different species from a common ancestor. Evidence can be found everywhere. For example, a species such as *Felis leo*, a lion, shares common traits with other members of the *Felis* genus, including *Felis pardus*, the leopard; *Felis tigris*, the tiger; *Felis silvestris*, the Scottish wild cat. Different species, yes, but all these cats have many things in common, including retractile claws,[12] and that's because they descended from the same ancestor. The same logic applies to human evolution, of course, though Darwin would not make that unpopular argument until thirteen years after he published *On the Origin of Species*. In *The Descent of Man*, first published in 1859, he explained how the morphological characteristics of the past become evident in present-day humans to "reveal the descent of man from some lower form in an unmistakable manner."[13]

All living forms descend from a common origin but are perpetually transformed through interaction with their habitats. The variations that adapt best to changing conditions survive. Science writer Olivia Judson describes the process with a clear example:

> Imagine you have a population of algae living for generations in a comfy freshwater pool. Now suppose there is a ghastly accident and, all of a sudden, the pool becomes super salty. Whether the algae will be able to survive depends on whether any individuals already have any capacity to survive and reproduce in salty water. If none of them do, they all die, and the population goes extinct. But if some do, then the survivors will reproduce, and over time beneficial mutations will accumulate such that the algae get better and better at living in a high-salt environment.[14]

In essence, nature does the same kind of work as do the breeders of domestic animals, just not so efficiently. The crucial point was now clear to Darwin: if breeders can create viable variations among animals in captivity, then in the wild and over a much longer stretch of time—what Darwin called the "long roll of years"—an immense and unstoppable diversity of species should appear. Any hint of God's meddling hand in the process had disappeared along the way. The awesome variety we see everywhere in the world today could have only been fashioned by a "blind watchmaker"—natural selection.[15]

Darwin had identified what has now become axiomatic in biological science: diversity sustains life. "The truth of the principle, that the greatest amount of life can be supported by great diversification of structure, is seen under many natural circumstances."[16] Living things evolve through uncharted processes that produce heterogeneous outcomes. "Slight changes in the conditions of life are beneficial to all living things," he reasoned, because "there will be a constant tendency in natural selection to preserve the most divergent offspring of any one species."[17] In the competition over resources, favorable variations will win out and can even become new species. Migration and adaptation determine what happens, according to Darwin. "New forms produced on large areas . . . will be those that will spread most widely, will give rise to most new varieties and species, and will thus play an important part in the changing history of the organic world."[18]

The classic though incomplete and somewhat misleading account of how Darwin developed the idea of organic speciation took place a thousand miles off the coast of Ecuador in the Atlantic Ocean. Stepping across the craggy topography of the Galápagos Islands nearly four years into his journey on the *Beagle*, the story goes, Darwin was impressed by the diversity he observed. He tried to understand what could account for noticeable differences in the size and shape of the beaks of mockingbirds and finches across the archipelago. Darwin noted that each beak seemed to be remarkably well adapted for the task of finding food on the island where the bird lived. Interaction with diverse environments was apparently changing the physical characteristics of the birds.

The gist of this well-known story is true, but what actually happened is not so straightforward. Darwin did not immediately grasp what had happened to the birds. He didn't even methodically collect specimens of birds on the Galápagos. The birds and lots of other organic samples he sent back to London had to be sorted out later, and some were lost for years. Many other experiences during the long voyage of the *Beagle* also influenced Darwin's thinking about how species come into being. He had been puzzled, for instance, by the fact that two species of the flightless ostrich-like birds called the Greater Rhea and the Lesser Rhea could be found in the same part of South America—Argentina.[19] Thousands of details had to be worked out. Darwin's major theoretical breakthroughs took place in the twenty-three-year period between the return of the *Beagle* to England in 1836 and publication of *On the Origin of Species*.

In keeping with the spirit of Darwinian mythology, the Galápagos birds did find their way into the narrative about evolution, and the role they played is important. Darwin presented samples of the birds he collected on the Galápagos to his friend, the renowned British ornithologist John Gould. The scientific questions the men addressed proved to be critical. Were the birds found on each island born as a separate species from the start? Or had evolution—much like the studied handiwork of a domestic breeder—transformed the original mockingbirds and finches by slowly changing their physical attributes in ways that increased their survival prospects in niche environments? If the birds were always of a separate species, that would align with the biblical interpretation of organic creation. But if the different-appearing birds were of the same species, then Darwin was describing how individuals of a species adapt to diverse environments, how they change, and, eventually, in some cases, how they become different species—biological evolution.

After studying the birds closely, Gould could not surrender his belief that the specimens represented different species. Although no definitive scientific proof could be summoned to decide the issue, Darwin had made up his mind in another direction. Scientific evidence now clearly confirms that Darwin was correct. The proposition Darwin put forward has become a biological principle: the wide variety of animals that is present all over the earth can only be explained by the process of long-term geographic speciation—the mutation of a species relocated from the natural environment from which it originated or from adaptations to major changes in that environment. In the case of the mockingbirds, finches, and huge tortoises of the Galápagos Islands, genetic adaptations over countless generations allowed the animals to feed themselves more easily in the diverse habitats in which they lived. Like those individual algae that adapted successfully to saltwater, nature "selected" the individual birds and tortoises with the right equipment.[20]

The discussion of natural selection outlined in *On the Origin of Species* and later in *The Descent of Man* has two implications that continue to generate unease and resistance today. First, Darwin posited that nature exhibits "one general law, leading to the advancement of all organic beings, namely, multiply, vary, let the strongest live and the weakest die."[21] No matter how sensible the idea may seem now, survival of the fittest still remains controversial—often for understandable and commendable reasons. Darwin never argued that natural selection is morally right, only that it's true. Second, he finally addressed the issue he carefully avoided at first: the same principles that explain change in animal evolution also apply to humans. Darwinian theory thus reflects what good science is all about: a material law is set into motion, and the rest falls into place.

The theory of natural selection insinuates a realistic, strictly secular view of biological processes consisting of two primary planes—inherited and environmental. Each plane has its own axis of possibilities but is also linked to the other. The *inherited legacy* reflects the genetic trajectories through which species adapt or fade away over time. Genes replicate and imperfectly transmit information that leads to complexity and diversity that is subsequently brought about by natural selection. In *The Origins of Life*, John Maynard Smith and Eörs Szathmáry describe how this happens: "What is transmitted from generation to generation is not the adult structure, but a list of instructions for making that structure. As fish evolved into amphibians, or reptiles into birds and mammals, the instructions changed, essentially by random mutation and selection."[22] Random genetic mutations make up the raw material for evolution, which is then acted on by nonrandom selection processes.[23]

Natural selection filters the variation because genes can be good or bad for the individual.[24] At the genetic level, survival is an (unconscious) group effort. As Richard Dawkins explains, natural selection "will see to it that gangs of mutually compatible genes will arise, each one selected for its capacity to cooperate with the others it is likely to meet in bodies, which means the other genes of the gene pool of the species."[25] That's how natural selection carves and whittles gene pools into shape.[26] Of special interest to us, for example, is the fact that within just the last five thousand to fifteen thousand years—the blink of an eye in evolutionary terms—some seven hundred regions of the human genome have been shaped and reshaped over and over again by natural selection.[27]

The second plane, *environmental interaction*, refers to how plants and animals evolve in the habitats they occupy. Progress results from the buildup of adjustments that are made from the bottom up in the self-interest of individual organisms interacting with other organisms and their environments. This image of life as a vast and vibrant succession of diverse interactions taking place in the natural world appears presciently in the first paragraph of the chapter on natural selection in *On the Origin of Species*: "Let it be borne in mind how infinitely complex and close-fitting are the mutual relations of all organic beings to each other and to their physical conditions of life."[28] Although biological change results from both heredity and environment, it's difficult to know how much influence to ascribe to either. Darwin admitted as much: "We cannot tell how much . . . to attribute to the accumulative action of natural selection, and how much to the conditions of life."[29]

The nature versus nurture debate had begun.

Without natural selection, organic life would expand limitlessly in all directions and bring catastrophic consequences. Even slow and parsimonious breeders like elephants would overburden their environments. But runaway growth doesn't happen. If predation, disease, and natural disasters don't reduce the growth of a species, then overpopulation will be curtailed by a gradual shortage of food. The individuals who triumph are the ones that adapt best to the demands of changing environments. The adaptations can be quite remarkable. Darwin illustrated this with the wonderful example of the flying squirrel. Some squirrels learned to fly because of the "changing conditions of life," which gradually led some of them to develop "fuller and fuller flank membranes" until the "accumulated effects of the process of natural selection [produced] a perfect, so-called flying squirrel."[30]

This death-defying process amounts to what is commonly called survival of the fittest. But close examination reveals that survival of the fittest is a hopelessly reductionist and circular idea. Which organisms survive? The fittest ones. How do we know they are the fittest? Because they survived. Obviously, an additional explanatory concept was needed.

APPROACHING SEXUAL SELECTION

Halfway through the chapter on natural selection in *On the Origin of Species*, Charles Darwin refers cautiously to sexual selection—the process of mating and reproduction—for the first time. Originally, Darwin thought the more encompassing idea of natural selection must be of greater importance because it necessarily implies death. At first, the repercussions of sexual selection seemed relatively minor compared to the terminal hardship inevitably brought on by natural selection. Gradually, however, Darwin came to believe that sexual selection is not at all a subordinate force.

While death resulting from the ravages of natural selection may seem to be the more dramatic outcome, non-procreation has determining evolutionary consequences, too. What good does it do for a tortoise to live a hundred years, for instance, if it doesn't reproduce? Realizing this, Darwin argued that natural selection must be interpreted together with sexual selection. Sexual selection, as Nicholas Wade puts it, "is a form of natural selection but one that works through mating success rather than physical survival."[31] Although Darwin didn't develop this crucial synthesis more fully until the second part of *The Descent of Man*, he ultimately gave sexual selection a central place in his system of ideas. He contrasted the selection processes this way: "Sexual selection depends on the success of certain individuals over others of the same sex, in relation to the propagation of the species; whilst natural selection depends on the success of both sexes, at all ages, in relation to the general conditions of life."[32] Natural and sexual selection thus function as a biological division of labor. The two processes intertwine: *sexual selection is the immediate, local method by which natural selection takes place.*

Some organisms reproduce in a simpler way. Asexual organisms multiply parsimoniously through direct copying—the splitting of identical genetic material. Significant sexual differences still don't exist today among such organisms as amoeba, yeast, and other fungi, flatworms, green algae, and other microscopic plants. By comparison, sexual reproduction is enormously burdensome, which is why Richard Dawkins understandably asks, "Why did sex, that bizarre perversion of straightforward replication, ever arise in the first place? What is the good of sex?"[33]

Sexual replication offers evolutionary advantages over asexual replication for four main reasons. First, sexually produced organisms evolve more rapidly. Second, sexual organisms produce more varied progeny, allowing them to adapt to environmental conditions better than asexual organisms, to generate fewer harmful mutations, and to repair their DNA more efficiently. Third, sexually produced organisms can fuse with each other so that a gene that is present in one cell can transfer easily to another cell.[34] And fourth, sex may incrementally promote biochemical protection against parasites that asexual reproduction does not provide.[35] The uncertainties and risks associated with sexual reproduction pay off in the long run because they increase the probabilities of greater genetic variation, which helps organisms adapt to the ever-changing environments they inhabit.[36]

Individual organisms want to survive, but that means much more than eating and avoiding being eaten. To avoid obliteration, the organism must also spread genetic traces of itself onto the next generation, forging another link in the evolutionary sequence. Males throughout the animal kingdom unconsciously seek out mates precisely for this purpose. But any individual male is less likely to be accepted by a mating partner if he is unable to communicate the required standard of fitness. By selecting outstanding mates, females determine which males' genes will be preserved in the next generation and which will not. No wonder males of all species will cheat, fight, deceive, and even give up their life for sex. Of course, the consequences don't affect only the male's genetic survival. Effective communication makes the difference in the quest for reproductive success for males *and* females. Sex binds males and females together because it perpetuates the genes of both.

We identify with and depend on our sex organs to such an extent that we sometimes develop personal relationships with them in ways we don't establish with other organs and appendages. A woman probably wouldn't have names for her arms, for example, but she or her partner might name her breasts. A man wouldn't call his right leg "Junior," but that's a common nickname for a man's penis. Men might even talk to their penis, and sometimes the penis talks back. Consider the content of a public service announcement that aired on Brazilian television. The nation's health minister was concerned that while most Brazilian men know the dangers of sexually transmitted diseases, they refuse to wear condoms. To encourage condom use, a public health campaign was launched. Here's the narrative from one of the spots, a conversation between a man and his penis, "Braulio." They're sitting together at a party:

> *Braulio:* "This place is full of interesting women!"
> *Man:* "Behave yourself, Braulio."
> *Braulio:* "How do you expect me to behave with so many beautiful women here?"
> *Man:* "OK, but if you come out, you're going to have to use a condom."
> *Braulio:* "OK, you win. But get the condom out quickly because there's a gorgeous woman staring at me!"

Brazilian health authorities called off the TV condom campaign, however, because of complaints from Braulios—men, not penises—from all over Brazil!

The claim that sexual selection serves as an evolutionary building block assumes two basic conditions. First, structural differences exist in the shape and size of the sexual organs of males and females, which themselves have been gradually adapted to fit together. These basic anatomical differences attract the opposite sex because neither gender can reproduce without access to the other's sexual organs. The unadorned, nude body communicates the difference. Sexual organs, especially the penis of males and the breasts of females, become "sexual signaling devices that serve to attract mates."[37] The physical attractiveness of the body impacts heterosexuals and homosexuals alike. We all engage in sexual practices that are inspired by the overwhelming instinct to procreate.

Second, we don't just evolve as a species; we evolve as males and as females with complementary social roles as mating partners. The basis of male-female pair bonding is simple and straightforward. Among hominids, as naturalists Donald Johanson and Blake Edgar point out, each sex had something to offer the other: "The male provides a reliable source of food as well as added protection for the female and the young. The female guarantees that the male's genes will be passed on to the next generation."[38] The biological history of our own species demonstrates the plausibility of this idea. The origin of hominid bipedalism can be traced in part to the fact that freeing up males' hands allowed them to gather and carry food back to females in the group. The males then exchanged the food for sex.[39]

Natural and sexual selection differ in terms of fitness—the determining evolutionary quality. In general, natural selection refers to the *overall* fitness of an entire species concerning environmental tolerance, disease resistance, resource utilization, and the ability to repel predators. But species are composed of individual members. This means that the overall fitness of any species refers to the average fitness of every one of those individuals. Each individual must be able to survive. The process of sexual selection thus refers to particular interactions between organisms among whom the level of *individual* fitness comes directly into play. Sexual selection, in Darwin's words, is "the process by which an individual gains reproductive advantage by being more attractive to individuals of the other sex."[40] That's where the selection processes interact. Natural selection produces characteristics that are selected sexually—the strong and socially directed male who carries food back to the female will be favored sexually, for example.[41]

TO CALL AND CHARM

The role of communication in natural selection revolves around the transmission of information; a mother duck helps her offspring survive by modeling how to swim, for example. But in sexual selection, the role of communication is to call and charm. If natural selection is about engineering, sexual selection is about art.[42]

When choosing a mate, the female of the species cannot directly inspect the genetic profiles of her male suitors in order to evaluate their candidacy. Instead, she will attempt to mate with those individuals that most effectively reveal qualities that promise to produce healthy offspring. A sexual organism selects messages. The quality of the genes is unconsciously inferred from the communicated information. Males attempt to display an impressive physical appearance, creativity and agility, social skills, or some combination of these characteristics to win favor in the most severe form of male peer competition—attracting females. Extravagance is rewarded. Poor communicators reproduce less.

A male's striking physical appearance alone can reap reproductive rewards. In the animal world, these genetically endowed attributes include the shine of the skin, quality of the fur, body heat given off, symmetric feathers, luxurious plumage, distinctive scent, richness of color—any and all traits and modes of adornment that can be displayed. Well-developed cheek pads on male orangutans win courtship battles, for example. Fish and salamanders display their brightest colors during breeding periods. Male giraffes with the longest necks get female attention. Even male flies, crickets, crabs, spiders, and grasshoppers display coloration, ornamentation, and sound that evolved for the purpose of attracting females.

Not all males can flaunt weapons of mass seduction like a massive rack of antlers, huge teeth and jaws, or a bright red chest to win potential mating partners. Charles Darwin wrote much about how the males of countless species try to charm females by their actions—energy displayed in a courtship dance, graceful movements of the body, or mellifluous song, for example. Why else would birds sing so enchantingly if not to advertise their robust nature? Darwin observed that female birds "are most excited by, or prefer pairing with, the more ornamented males, or those who are the best songsters, or play the best antics."[43] Male hummingbirds attract females by producing high-frequency vibrations from their tail feathers, for example. Indeed, the pop music lyric "shake your tail feather," made famous by Ike and Tina Turner, the Blues Brothers, the Cheetah Girls, Nelly, and many other musicians, traces its roots not just to the blues but to biology. With their clothing, sounds, and movement, humans often imitate animals and birds for sexual reasons. In some cultures, sexual expression is influenced by the strong vocalizations and distinctive dance of the whooping crane, for example.

Tail feathers aren't the only body parts that birds and other animals shake. As Darwin described it, birds also display "combs, wattles, protuberances, horns, air-distended sacks, top knots, and plumes" for sexual purposes. Male crocodiles go "splashing and roaring in the midst of a lagoon [while] emitting a musky odor."[44] Male lizards do deep push-ups to demonstrate fitness. One species of spider choreographs a multimedia show of song and dance. Hedgehogs dance around prospective mating partners for days. Some birds tap out messages of seduction with their beaks, and rain forest monkeys show off their strength, cleverness, and agility by drumming frantically with sticks.[45] The red crowned crane throws sticks to impress females.

The timing and rate of physical movement, frequency and volume of sounds and calls, and persistence in the face of rejection can all lead to reproductive success in the animal world. "Family man" social skills on the part of desiring males can also win sexual favors. Grooming and sharing food are common tactics. Warning the female of danger can also be effective, even when done deceptively. When male topi antelopes in Kenya sense that a sexually receptive female is about to leave their territory, they strike a frozen stance and snort a warning that predators are nearby, even when they aren't. The female typically retreats, and the male attempts to mate with her before she leaves again. Female topi repeat this behavior many times in their lives because the risk associated with predation is so great.[46] Deception for sex occurs frequently in nature. Male nursery web spiders obtain sex by making gifts to females, like insects wrapped in silk. But a male may also package fake gifts (usually a piece of flower wrapped in silk) that allows him to distract the female long enough for his purposes before she figures it out.[47]

Monogamy is rare in nature.[48] The vast majority of males never win a female's permanent affection or loyalty. The pressure males feel to communicate and copulate is enormous. They not only have to impress potential mating partners, but they also have to stave off other males with whom they compete for the privilege. Throughout the animal world, female choice dominates sex and reproduction.[49] Richard Dawkins calls this natural phenomenon "the selective breeding by females of males."[50] The magnificent beauty, variety, and ornamentation of the animal world represent the net result of males' frenzied attempts to attract potential sexual partners. Pervasive polygamy and fierce sexual competition drive sexual dimorphism—the development of contrasting physical characteristics of male and female organisms that belong to the same species.

When females refuse to mate, males typically escalate the intensity of their sexual displays.[51] The inflation of sexual signals decorates the palace of organic evolution, accumulating over time to mark sharp physical differences between the genders. The breathtaking colors of male birds and polyphony of their songs stand in stark contrast to females' smaller size, less brilliant coloring, and lack of interest in singing, for example. Other physical and expressive characteristics of males—longer tails, vigorous antics, and louder songs—also evolve in response to female preference.[52] Conversely, those species among which males and females look and behave similarly are the most monogamous.[53]

To initiate a mating ritual, females characteristically send messages of sexual availability, often just by their conspicuous presence, which can put them at high risk from predators. The male responds with a message conveying his biological fitness. The female turns the message she receives into a meaningful sign through inference: if the sender makes an impressive display, he must be fit. Females select mating partners by appraising the totality of messages they receive. Often overwhelmed by suitors, females have to be discerning interpreters of the sexual signals sent their way and able to resist male actors' sales pitches to weed out the losers.[54] In some species, females signal acceptance or rejection before consummating sex. In the courtship ritual of common house finches, for example, males initiate the interaction by touching bills with a female and then presenting her with bits of food. She signals acceptance of the offer by mimicking the behavior of a hungry chick.[55] Each bird sends signals reflecting his or her survival needs to consummate the sexual act.

Sexual selection can be a very dangerous game. Effective sexual communication should cost as little as possible to message senders and receivers because no organism wants to attract the attention of predators. Thus, a trade-off between selection processes is evident: over time, natural selection forms organisms in ways that blend in with their habitats so they will not be detected by predators; sexual selection, on the other hand, produces flamboyant shapes and colors in males to attract mating partners.[56] This natural balancing act explains why males become more physically conspicuous in environments where predation is weak or nonexistent. Less predation inspires more expression.[57]

But prudence doesn't always serve the best interests of the senders or receivers of sexual messages. Despite the hazards of standing out, males don't want to conceal their special qualities, and females don't want the message senders to hide their talents. The male peacock's display of his impressive tail plumage may be risky, for example, but it demonstrates biological fitness in a most impressive way. Fitness gradations can be very subtle. For example, when dancers perform—whether they are honeybees, blue jays, or teenagers—small variations in body movement finely distinguish the best candidates for sexual selection.

HUMAN SEX AND EXPRESSION

In one of the most astonishing images from sports ever to light up television screens around the world, Zinedine Zidane, the famed former striker for the French national soccer team, suddenly head-butted the chest of Italian defender Marco Materazzi in the waning moments of the World Cup championship game in 2006. The blow knocked Materazzi to the ground and sent Zidane to the sidelines, where referee Horacio Elizondo issued the livid Frenchman a disqualifying red card. Without the powerful Zidane available for the penalty shoot-out of a match that had ended in a 1-1 tie, Italy took the Cup with a 5-3 win over a very disoriented and discouraged French team.

What could have caused Zidane to commit such an irrational act that he knew would result in a blatant foul in the world championship match? No one but Zidane and Materazzi know for sure what happened. Not surprisingly, the players' stories differ greatly. What all accounts have in common, though, is that Materazzi had been taunting Zidane throughout the match about the women in his family. One version has the Italian calling Zidane, who is of Algerian descent, the "son of a terrorist whore." Another says that during the match, Zidane contemptuously offered to give Materazzi his jersey after the game, a soccer ritual. The Italian responded by saying he'd rather have Zidane's sister. Just as mountain goats, elk, deer, and hippopotamuses head-butt each other to challenge or protect sexual territory and property, the Frenchman resorted to brute force to defend the honor of his family. The mighty head-butt quickly restored biological equilibrium and Zidane's masculinity—World Cup championship be damned!

Like alpha male primates, men sometimes beat their chests to show dominance. They taunt, threaten, and physically engage each other, sometimes violently, even fighting to the death. But most men and animals negotiate sexual selection less savagely. They rarely settle conflicts over access to females by fighting outright, preferring to signal victory in other ways.[58] Deer lock antlers and bend their knees, for instance. Snakes refrain from using their venomous fangs, instead ritualistically shoving their entangled bodies around on the ground until one concedes defeat. Fish grasp each others' jaws, pushing and pulling each other back and forth. The rules of conventionalized fighting reduce injuries by limiting physical contact. All contenders desire reproductive success, but not at the expense of their physical integrity. The limited war strategy benefits individuals competing to be sexually selected.[59]

Opponents in struggles over sex transform violent behavior into ritualized combat that is composed of an exchange of signs. Over countless generations, animals have created rule-based communication systems for signaling competitive outcomes like this. The weaker animal backs off. The stronger animal doesn't continue the fight because he has already won. The message "I'm better than you" has been effectively delivered, so there is no need for the victor to waste time and energy or incur physical injury—all evolutionary costs.

Even for humans—a very recent arrival on the evolutionary scene—sexual communication has a long history. The first personal symbolic forms fashioned by humans—the perforated shell necklaces discovered in Africa and the Levant—must have surfaced in response to the competitiveness of sexual selection. Ever since then, just like shells turned into jewelry, many of nature's resources have been adorned and transformed into material commodities that reinforce and advance reproductive instincts. Modern industrial capitalism has precipitously expanded the range of available material and symbolic resources used for sexual signaling. Consumerism in general can be interpreted as the construction of personal displays motivated by sexual selection.[60] Why would people buy fancy clothes, use makeup, wear jewelry, or undergo cosmetic plastic surgery, for example, if not to enhance their beauty or delay aging, thereby promoting and extending the potential for biological reproduction? Conspicuous consumption and waste transforms the consumer into a well-decorated and confident agent of sexual expression.

Men have always employed a variety of powerful images as

sexual credentials. The warrior stereotype—which includes soldiers, gang members, professional wrestlers, and gangsta rappers—endures today as an aggressive statement of sexual capacity. Cultural artifacts and behaviors—guns, tattoos, vocal threats, profanity, hard-edged music, fast and loud cars, death and devil imagery, for example—all function as masculinity-enhancement props. A particularly dramatic example is the display of animal body parts, especially the heads and skins of big beasts slain on safaris and other hunting trips, to demonstrate male predatory qualifications—animal and sexual.

As with all animals, the sexual success of humans depends on their ability to communicate effectively. Communication skill itself seduces. When Silvio Berlusconi was accused of paying prostitutes to service his sexual desires at all-night parties in his home, the rowdy Italian politician took great offense. Berlusconi was not outraged for being accused of cavorting with beautiful young women; he had a different reason: "I have never understood what satisfaction there is if not the pleasure of the conquest." Flirting and seduction is artful communication that promises the maximum payoff—sexual gratification and potential reproduction. Social skills—nuance, sensitivity, humor, intelligence—are revealed in the playful interaction. Flirters define themselves and learn about the other during the courtship stage—an important evolutionary component of this unique form of communication, because humans require an immense investment of time for pregnancy and parenting.

The contrasting pitch, timbre, and volume of the typical male and female voice intensify sexual attraction. The stereotypical deeper, more resonant male voice connotes confidence and authority—read sexual confidence and authority (the Barry White phenomenon)—a good bet for successful mating and the propagation of one's genes. The softer female voice signifies "not male," a reliable indication of childbearing potential. The sound of the voice plays out so importantly in sexual relations that popular instructional books like Bonnie Gabriel's *The Fine Art of Erotic Talk* have been written to provide instructions for creating effective vocal foreplay.[61]

LANGUAGE OF LOVE

Command of language must be in place before the individual can select or be selected sexually. Just as biological evolution proceeds from the simple to the complex, language development moves from the absence of understanding to complex comprehension. Composed of a limited number of phonetic units, languages are rich when compared to the impoverished and repetitive (yet precise and necessary) modes of communication used by animals. Infinite possibilities for innovation, recombination, and outcrossing function at the hub of linguistic systems in ways that resemble the work of information-passing genetic structures inside the bodies of living organisms. The value of language production for obtaining sexual success encourages humans to constantly add words and devise ways of speaking to their repertoires.

This ability to create, combine, and recombine linguistic resources, like the use of sampling techniques in music or video production, demonstrates to potential sexual partners that the individual can respond to challenges with innovative solutions and can make discerning decisions—signs of reproductive fitness. Research confirms what casual observers have thought all along—creative types attract more sexual partners than their less imaginative peers.[62] Writers, artists, journalists, broadcasters, film stars, and musicians appeal because they have proven to be good communicators of ideas and feelings—a favorable personal quality in the game of sexual selection.[63] A vivid imagination, creativity, expressiveness, artistic talent, stylish self-presentation, and celebrity are all qualities that individuals find personally attractive and desirable to pass on to their progeny.[64]

Intuitively, music seems to be an expressive extension of spoken language. Charles Darwin saw it the other way around. He believed simple musical expression led to the development of articulate speech and language. Competitive sexual selection was behind Darwin's thinking. Darwin thought that "musical tones and rhythm were used by our half-human ancestors, during the season of courtship, when all animals are excited not only by love, but also by the strong passions of jealousy, rivalry, and triumph."[65] Whatever its true origin, music communication connotes differences between males and females. The consequences are meaningful. If Neanderthal women really did sing to their children, as archaeology educator Steven Mithen speculates, that would have reinforced their role as females in a species not known for gender-divided labor.[66] The misogynistic male rock star or rapper stereotype may represent the most obvious examples of pure gender-differentiated sexual communication today.

The undulating movements of popular dance suggest sexual potential while mimicking the physical acts of copulation and childbirth. Observing someone dance gives clues about the person's potential as a sexual performer, suggesting the degree of procreative capacity.[67] Ballet—one of the world's most exquisite displays of the human physical potential—elevates sexual communication to high culture. In ballet, we see the normal limits of bodily performance exceeded and recontextualized for public consumption. The same can be said for competitive sports. Enormous audiences gather at arenas and around television sets to appreciate the spectacle of superior bodily performance. Male sports figures receive most of this attention as they show off physical qualities that women admire—size, speed, power, creativity, agility—attributes that better ensure pregnancy and represent desirable traits to pass on to progeny. Throw in the good looks of a David Beckham, Kobe Bryant, or Rafael Nadal, and you have an irresistible candidate for sexual selection—at least in evolutionary fantasyland.

Sexual selection for animals and humans remains an indirect inference, an informed guess rendered from communicated clues that may or may not reflect the best possible choice. The selector works out courses of action or inaction by assessing available mates. Sexual selection operates essentially the same way for nearly all advanced species: females choose potential partners from a consortium of available males who send messages of biological fitness and personal excellence. Subtleties vary from individual to individual, but the content of the male's urgent plea never wavers: "I am an exceptional organism. Copulate with me, and you will have fabulous offspring!"

Biology and culture make human sexuality and reproduction more complex than those of other species. Unlike the females of most other advanced species, women do not exhibit outward signs of ovulation, making it impossible for either gender to know for sure when she is most likely to become pregnant. For our hominid ancestors, this meant the males didn't have to physically fight over females who were obviously ovulating. But it did provoke other challenges. The inscrutability of hidden ovulation put pressure on males to become relentlessly devoted to sexual activity in order to reproduce.[68] Frequent sexual activity, the quality of sexual performance, and the robust production of offspring began to define human masculinity.[69] Reinforced by media and popular culture, humans have become sexually obsessed. Jealous rage drives some men to commit acts of domestic violence; some of whom go so far as to murder a competing male or a real or suspected disloyal female. Many men who cannot express themselves sexually become severely depressed.

Other social evolutionary traits also developed. Greater cooperation and improved communication between males and females became increasingly adaptive. A constantly present male became an important evolutionary asset. Males not only performed a sexual function for females and themselves, but they also participated in a rare behavior among animals but central to mammals—parental care. Cooperative, communicating males and females developed reciprocal traits that ranged from nurturing family alliances to trading food for sex, an exchange that is also present among other primates.[70] Functional interdependency that developed between the sexes does not mean humans became naturally monogamous, however, as contemporary trends make abundantly clear.

Many religious leaders and their devoted followers fight the intense urges that excite sexual behavior throughout history, admonishing against what, to them, are obvious sins: intercourse before marriage, infidelity, and homosexuality. Religious authorities also try to control the politics of sex: abortion, stem-cell research, and the legal terms of marriage and consequences of divorce, for example. Yet despite the stern warnings, threats, and political influence, fewer couples are getting married now, more people have sex outside marriage, and divorce is on the rise even in the strictest societies.[71] Media and cultural globalization encourage people to act in less restrictive ways.

In summary, five factors govern most evolutionary outcomes: (1) *random mutations* are sorted by processes of (2) *natural selection* and (3) *sexual selection* that take place in (4) *changing environments* (5) *over time*. Natural and sexual selection aren't the only processes that influence biological change, but they are by far the most powerful. The random production of mutations can also create genetic drift—heritable variation that shapes a population by chance, not by selection. And some causes of change are simply unknown. This is not news, nor does this fact diminish the explanatory power of evolutionary theory. The likelihood that some mutational spin-offs lead to genetic drift—a condition most often present in isolated populations—was recognized by Darwin long ago and has been documented, discussed, and debated by many evolutionary theorists ever since.[72]

DARWIN'S DILEMMA

When Charles Darwin went public with the idea of natural selection, he described it almost as a metaphor in hopes that a slightly abstract approach might make evolution seem less strange and threatening. But the implications were difficult to ignore. If artificial selection allows the breeders of dogs, pigeons, and plants to tinker successfully with the natural order under controlled conditions, then the same principle must apply to the unsupervised realms of nature as well. As Darwin put it, "I can see no limit to this power, in slowly and beautifully adapting each form to the most complex relations of life. The theory of natural selection, even if we look no farther than this, seems to be in the highest degree probable."[73]

But Darwin did look farther, by integrating sexual selection into natural selection and giving it a prominent place in his comprehensive canvas of evolutionary theory. As always, Darwin presented his ideas carefully. Describing a decisive evolutionary role for sexual selection at the time he wrote *On the Origin of Species* would have created great difficulties. Naturalist Alfred Russell Wallace, who had come to the idea of natural selection simultaneously but independently from Darwin, repudiated the notion of sexual selection. Other authorities were likewise unconvinced. Darwin thoughtfully organized the main argument in *On the Origin of Species* around natural selection, not sexual selection, and avoided any discussion of how humans evolved. He didn't begin to tell the more complete story of humans' development for twelve more years, when he described sexual selection in *The Descent of Man*.

Recognizing the importance of sexual selection represents one of the most important steps in the maturation of Darwin's system of ideas. The connection between the bottom-up nature of human evolution and the power of social communication is fully evident in sexual selection. Sexual selection makes mutation a matter of the individual; by choosing mating partners, individuals play a central role in forming future generations. More than a century after Darwin first explained how selection processes shape organic evolution, his thinking about the main trajectories of evolution has become axiomatic: natural selection reflects avoidance of the end of life; sexual selection manifests life's beginning.

CHAPTER 4

COMMUNICATING CULTURE

Published toward the end of last century, Samuel P. Huntington's *The Clash of Civilizations and the Remaking of World Order* stirred up a hornet's nest that is still buzzing and sometimes stings. But Huntington's controversial argument—that many cultural groups have become fundamentally incompatible—seems downright mild compared to Charles Darwin's claim many years before that the world can be divided between "civilized and barbarous races."[1] Huntington demarcated cultural groupings and identities in terms of the complexities of post–World War II geopolitics. Darwin wasn't even sure at first that all races belong to the same species.

Without genetic data or much of a fossil record to guide scientific thinking in the nineteenth century, the idea of geographically distributed speciation was a reasonable way to account for why humans of different continental origins vary in physical appearance and cultural behavior. How else could Darwin explain the morphological contrasts and diverse ways of living he observed between upper-middle-class families of Victorian England and the aboriginal cultures of South America and the South Pacific? After all, Darwin surmised, *Homo sapiens* is just the latest of many hominid species to appear on the earth. Other species may still exist and more will likely follow.

Presenting massive empirical evidence he subjected to rigorous examination, Darwin described in *On the Origin of Species* how "forms of life throughout the universe become divided into groups [that are] subordinate to [other] groups."[2] He was trying to solve a puzzle. From the time he was a boy collecting beetles, the nagging matter of how to classify the natural world stimulated Darwin's imagination about evolution. After studying nature carefully; reflecting privately and deliberating with others; and then studying, reflecting, and deliberating over and over, Darwin formulated a powerful yet parsimonious description of how biological species come into being: common descent with modification. Scientific research confirms Darwin's theory. Organic life began with a common ancestor—a unicellular organism that emerged in the primeval soup three and half billion years ago. All organisms change as they pass through time and eventually become separate species. At every stage, the organisms develop complex relationships within their species, with other species, and with their physical surroundings—which are also all in flux.

If common descent with modification accurately describes the proliferation of species—and it does—then the various ethnic and cultural groups that populate the world represent *communities of descent*. We share a common biological inheritance, but our physical mobility has led us to populate even the most remote places on the planet. Selection pressure and the biological requirements of reproduction forced individuals into tight patterns of interdependency wherever they went. Our ancestors grouped together to escape death and to compete for territory and resources. Accumulated knowledge, customary behavior, and technical skills adapted to various living environments began to differentiate one group from another.[3]

Cultural groups tend to remain cohesive because their members believe the collectivities to which they belong serve them well.[4] The sense of belonging and identity that culture provides becomes profoundly associated with the individual's perceived chances for survival. With so much at stake, tribal affiliations and allegiances have become deeply engrained in the human psyche. Penalties that would be suffered for abandoning one cultural group for another are often prohibitive, sometimes unthinkable; even today, apostasy can be punishable by death.

Human communication skills developed in tandem with harmonious group behavior. Unlike all other species, we have created elaborate, culturally specific codes for social interaction that provide extraordinary advantages for cultural groups. Sharing information and telling stories in the local language nurture a sense of common interest, encourage and reinforce cooperative reasoning, teach expected behavior, help individuals gain social acceptance, and protect the group from interlopers.[5] Communication errors identify outsiders. As Michael Tomasello puts it, "Anyone who does not speak our language is not one of us, but also anyone who does not dress like us, or eat like, or paint his face like us, or worship like us, or all kinds of other things."[6]

Thousands of years of migratory behavior may have set us against each other, but the fundamental nature of culture and communication offers great hope. Survival depends as much on cooperation as it does on competition, as much on inclusion as exclusion. Humans are problem solvers. Our power to purposefully change the world for the better developed as a by-product of the brain's capacity to meet practical challenges, a capacity that is facilitated by language.[7] Our ability to process complex information, to imagine diverse courses of action, and to create alternatives to the way things currently exist separates us from other life-forms. The cooperative nature of our species leads to unending development of our communication ability, which then continues to positively shape the way we live together.[8]

GENES, CULTURE

Genes replicate. People communicate. Genes don't think about what they're doing. People presumably do. Nature and culture do not make up separate spheres of life; culture stems from nature. But biological change differs from cultural change in fundamental and important ways. Biological replication is comparatively direct. Among sexual species, genetic information passes from parent to offspring as sequences of DNA. Transmission takes place only once—at the moment of conception. Once the seed is planted, nature does the rest of the work. Success of the organism depends largely on the physical tolerance and nurturing qualities of the parents, especially females, and sheer good luck. Biological evolution has no goals or interest in human welfare.[9] Although errors in replication occur, genetic information transfers efficiently from one generation to the next. Cultural transmission, on the other hand, requires that useful information external to biological organisms passes from one generation to the next. The process is far less straightforward than biological evolution and is much more open to variable outcomes, even to outright resistance and rejection. While biological mutations are random and selected only after they occur, cultural choices are purposeful and motivated from the beginning.

Our biological selves are not determined by genes, and our cultural selves are not dictated by history and tradition. Genes and culture travel an uncharted course together, adapting to fluctuating environments in a never-ending process of biocultural feedback.[10] As anthropologist Clifford Geertz describes it, "Between the cultural pattern, the body, and the brain, a positive feedback system was created in which each shaped progress of the other."[11] Evolution is inherently synergistic. Genes make up stretches of DNA that contain instructions for making protein molecules, functioning like a recipe for biological growth and behavior.[12] The ingredients and cooking process, however, are set by the material and social environment.[13]

This gene-culture/culture-gene interplay may even help explain one of the great mysteries of human behavior: how the psychological dispositions of individuals and the internal cultural patterns of groups develop, shape consciousness, and prompt people to act in particular ways.[14] Human behavior has shifted over time from situation-specific observational learning to inborn knowledge that is stored in the brain by means of genetic assimilation learning.[15] Social interaction then gradually modifies the architecture of the brain. Behaviors that afford a fitness advantage will be selected and eventually wired into the neural circuitry of future generations. The resulting genetic structure reflects the inculcated behavior and helps direct, but never fully determines, its future deployment.

Throughout history, large and dramatic shifts in biological evolution have evolved as a result of weak forces operating over vast periods of time.[16] Lately, however, the pace of biological evolution has been quickening. The Human Genome Project has produced convincing evidence that the biological evolution of our species has been speeding up rapidly since behaviorally modern humans first began to appear about fifty thousand years ago.[17] In just the past ten thousand years, the pace of biological evolution has increased especially quickly compared to any other period since the hominid line split off from chimpanzees seven to five million years ago.[18]

The parallel pace of biological and cultural change evident since our ancestors left Africa is also striking. Sedentary farming, which began to replace hunting and gathering about ten thousand years ago, dramatically changed cultural life, including dietary habits, which led to historically rapid genetic changes.[19] Scientists point to two recent research projects to show how cultural behavior can affect the genetic code. In the first case, people living on the Tibetan Plateau have adapted to the extremely high altitude there within just the past three thousand years. Populations that have not made this genetic adaptation suffer serious health problems in the high altitude.[20] In the second case, immunity to lactose disorder develops in cultures where people drink milk throughout their entire lives, a trait associated with dependence on dairy farming. Over the past 7,500 years, this adaptation has created an enzyme that safeguards the digestive system and allows the cultural practice to continue.[21] Other populations don't exhibit the same immunity.

Genetic evolution also speeds up with technological development—especially the growth of communications technology—although the links between these kinds of cultural transformations and biological change are more difficult to demonstrate empirically. What is clear, however, is that culture is changing much faster than ever before. As the rate of information transmission increases, so, too, does the degree of cultural complexity. Major cultural transformations are taking place worldwide today as the mass media and culture industries function together with information and communications technologies to incessantly transmit big ideas to enormous audiences with unpredictable, sometimes uncontrollable, consequences.

FROM GENES TO MEMES

More than thirty years ago, Richard Dawkins put forward the idea that organic life's fundamental hereditary unit is the key for understanding evolution's long-term outcomes: "They are in you and in me; they created us, body and mind, and their preservation is the ultimate rationale for our existence. They have come a long way, those replicators. Now they go by the name of genes, and we are their survival machines."[22] Rather than consider human evolution from the point of view of the organisms or groups that have survived, Dawkins says, we should think of evolution from the perspective of the genes that have been passed on from generation to generation. But he also makes it clear that genes don't explain all of human evolution. As Dawkins described the power of the "selfish gene" in a book by the same name, he also introduced the "meme" as a way to conceptualize the presence, appeal, and spread of enduring cultural themes and traits.[23] Drawing illustrative parallels between genetic replication and cultural, or "memetic," transmission, Dawkins speculated about how some cultural ideas establish and maintain their strong influence: "Just as genes propagate themselves in the gene pool by leaping from body to body via sperms or eggs, so memes propagate themselves in the meme pool by leaping from brain to brain via a process which, in the broad sense, can be called imitation."[24]

The *Oxford English Dictionary* defines *meme* as an element of culture that may be passed on by nongenetic means, especially imitation. It is cultural information that replicates and propagates itself.

Social scientists refer to memes when describing the circulation and impact of cultural phenomena. A growing number of media pundits use the term when talking about the explosion of cultural trends.

The meme is a succinct, catchy, and sensible descriptor of a huge and unwieldy idea. Its attractiveness as an explanatory concept, however, can mislead. Dawkins himself never claimed that memes resemble genes precisely, or that meme theory adequately explains cultural transmission: "I am not saying that memes necessarily *are* close analogues of genes," Dawkins explains, "only that the more like genes they are, the better will meme theory work."[25] He proposed the meme as a medium that is capable of spreading ideas and producing cultural patterns but made it clear that he "never wanted to push [memes] as a theory of human culture [but] almost as an anti-gene point—to make the point that Darwinism requires accurate replicators with phenotypic power, but they don't necessarily have to be genes."[26]

Despite Dawkins's warnings, the meme itself has become a meme. So what's at work here? How do memes come into being, travel, enter individual minds, and affect consciousness?

American philosopher Daniel Dennett believes that the passing of cultural ideas from one person to another corresponds methodologically to the way multicellular organisms first came into being.[27] Although scientists have yet to explain with great certainty how multicellular organisms first appeared on the earth, the best guess is that various parasites invaded and inhabited unicellular organisms as they reproduced asexually in the primeval soup. This merging of unicellular organisms gave rise to biological symbiosis and engendered the first signs of multicellular life. Something similar may be happening in cultural transmission. A process of assimilation enables the transfer of cultural elements from one person to another, leading to the creation of cultural themes and social norms. In Dennett's words, the meme is "a data structure with attitude."[28]

From the songs you can't stop singing in your head to fantasies dreamed up about heaven and hell, some cultural elements manifest great impact and staying power. English psychologist Susan Blackmore argues that entire modern cultures represent "the legacy of thousands of years of memetic evolution."[29] Memes "are instructions for carrying out behavior, stored in brains," she says, much like "genes are instructions for making proteins, stored in the cells of the body."[30] Blackmore focuses on the causal mechanism of memetic transmission emphasized by Dawkins—social imitation. Learning how to copy each other's actions gave early humans superior ways to think and set the stage for later cultural transmission and development. The process amounts to straightforward social communication and iterative learning (driven by immediate feedback). An imitator-sender encodes a message; a receiver-imitator decodes one. The meme facilitates cultural transmission and functions as a unit of shared cultural meaning—complementary roles that actualize at the moment individuals interact.

Ranging from the material to the abstract, memes reside in everything around us. A meme can refer to a tiny bit of material culture—a sauce used for cooking, for example—or it can represent the least physically tangible dimension of culture—the notion of a personal God. The idea of a vegetable is a meme, and so is vegetarianism. An aluminum can is a meme, and so is recycling. A linen blouse is a meme, and so is fashion. Some of the ideas in our worlds and in our heads (sauces, gods, vegetables, vegetarianism, cans, recycling, blouses, and fashion) get copied with such frequency that they come into high relief and persist over time.

A reasonable criticism of the meme as an analytical concept is that it represents nothing more than the mere identification of hot cultural trends or lasting themes. It can also be argued that the spread of cultural ideas can be explained by the dynamics of social contagion theory—emotional and behavioral copying that flows mysteriously from person to person.[31] But these objections fail to recognize the special characteristics of memes. Memes perform as uniquely commanding motivators of cultural transmission because they are free-floating elements. Memes take on lives of their own.[32] The meme, therefore, can properly be described only with a compound definition: *memes are cultural ideas that inhabit the minds of individual human beings who pass the ideas along to others, but they also exist independently from their human hosts.* Some memes survive at the expense of other memes. In keeping with Dawkins's classic formulation about genetic transmission, the ones that live on could rightly be called "selfish memes."[33]

Like genes that require a home base from which to operate—the double-helix configuration of DNA nested within a biological carrier—memes need help, too. Cultural transmission requires a continuous exchange between articulating agents and the cultural milieu. Terms like *carrier*, *vehicle*, or *medium* seem too one-directional for the job. *Interactor* expresses the idea better.[34] The most common and useful interactors are people. People acquire, embody, transport, communicate, and give credibility to the cultural materials and ideas they host. The transfer of cultural information is by no means limited to face-to-face interpersonal communication. Mass media, personal communications technology, and the Internet have become particularly potent transmitters of cultural information.

Just as the success of a gene is associated with the success of the cell and of other genes in the cell, memes also act in concert. Dawkins refers to these mutually reinforcing cultural associations as a "memeplex."[35] For example, the automobile belongs to a memeplex that contains powerful cultural ideas having to do with general concepts of machinery, transportation, freedom, responsibility, style, licensing, and regulation, as well as with specific brands, logos, advertising campaigns, and so on. Furthermore, today's automobile descends from invention of the wheel, discovery of the axle, fusion of the drive train with the combustion engine, industrialization of the assembly line, refining of petroleum to produce gasoline, and many other identifiable cultural phenomena.

We recognize the telephone as personal communications technology. But the telephone was designed originally as an aid for the hearing impaired and later became a surveillance and information tool used during wartime. Each of these cultural moments is still part of the modern telephone today. The telephone continues to serve as a hearing aid, an essential piece of military equipment, and an information-sharing, consumer-driven communications medium facilitated by fiber optics, satellite technology, and the global telecommunications industry. An evolutionary principle takes effect here: remnants of foregoing types like those present in the automobile and telephone also persist in the biological world. Darwin's law of reversion describes how the "long lost character" of previous generations appears in the offspring of animals many generations down the line.[36]

Despite appearances to the contrary, neither genes nor memes have intentions or take initiative. Still, they evolve in ways that are advantageous to themselves, or they wouldn't exist. Like the healthy diversity that arises within and among the most complex biological organisms, memes often become good survivors when they contribute to the composition of complicated belief systems.[37] Selection principles influence the social and technological aspects of cultural transmission just as they shape biological proliferation. The fittest memes—those that attract the most attention, good or bad, and maintain their cultural presence—are the ones that affect the circulation of cultural ideas. Evolutionary benefits accrue to the messenger as well: the best human communicators of cultural information are the ones most likely to survive, reproduce, and pass along their genes.[38]

Culture is an extremely nebulous concept. The constituent elements of culture are similarly confounding. No one has ever seen a meme. But that fact shouldn't surprise or discourage. No one had seen a gene either until DNA was discovered in the middle of the twentieth century. The abstract concept of the gene had nonetheless been anticipated more than a hundred years earlier when Augustinian monk Gregor Mendel studied the inherited qualities of pea plants. Mendel believed that some principle of biological particularity must underlie the organic inheritance of plants and probably of animals, too. Genes later became identified as that biological agent, and genetics developed as the system that explains the inheritance of living organisms. So far, we don't have the same quality of insights or scientific evidence to explain how and why certain cultural traits develop and endure while others do not. We're too often left with the dubious explanatory power of an imprecise concept hooked to a misleading metaphor—the meme as a virus, for instance.

THE VIRULENT MEME?

How memes and memeplexes spread their influence has been compared to the way a virus can ruin a body or machine. From this point of view, memes infect our lives. They are contagious. Parasitic. Memes invade worlds and control how we live. There is no escape. In *The Meme Machine*, Susan Blackmore argues that the cognitive algorithms of memes render cultural freedom illusory. Consciousness doesn't exist apart from the tyranny of memes because a totalizing homology develops between the agents of memetic reproduction and the human mind. We humans have no independent mind to protect ourselves from "alien and dangerous memes."[39]

Is this true? Are we such passive casualties of culture?

Cultural concepts, traditions, and practices frame and guide our thinking, often subconsciously, and not always to our benefit. They socialize us to accept cultural assumptions and to conform to social expectations, usually without questioning their history or whose interests they represent. Winners and losers emerge in social relations that play out culturally. But the meme-as-virus metaphor fails to adequately explain how cultural transmission actually takes place and what the true consequences are. Human beings have not evolved through the millennia only to become inert repositories of cultural ideas that fix their values and direct their behavior. Viruses bring only misery. Cultural life, even under repressive conditions, is not determined in the same way viral infections ravage biological bodies or destroy computer networks.

Charles Darwin thought about how ideas spread, but he focused on what he considered to be the essentially positive nature of the process. In *The Descent of Man*, Darwin wrote of how members of various cultural groups could eventually overcome their differences and begin to "look at [people who make up other groups] as our fellow creatures."[40] Communicative interaction would make this possible: "As soon as [concern for the welfare of others] is honored and practiced by some men, it spreads through instruction and example to the young, and eventually becomes incorporated into public opinion." Darwin was describing the attractiveness of life-affirming good ideas—in essence, the liberating influence of soft power.

Darwin's observations provide necessary perspective. Comparing cultural transmission to the spread of a virus brings theories of cultural imperialism to mind. The regrettable story is well-known. For centuries, the imperial nations exploited the less developed parts of the world, especially Africa, Asia, and the Americas. They ransacked the colonies for economic and geopolitical gain. Indigenous cultures were plundered. Even decades after the colonized nations won their independence, political, economic, and cultural influence continues to emanate from England, France, Spain, Portugal, the Netherlands, Russia, Japan—and especially from the United States. Drawing intellectually from classical Marxism and from ideas about a defenseless "mass society," cultural imperialism theory follows the same line of reasoning. Big media and transnational culture industries corrupt local traditions, pollute consciousness, manipulate behavior, and undermine the human potential.

Cultural imperialism theory draws from classical humanistic philosophy and liberal politics. It merits respect. The basic Marxist argument, simple as it is, will always bear an element of truth. But the technological landscape has changed so much in recent years that arguments based on last century's cultural realities no longer advance the discussion productively. The diversified production and flow of symbolic resources taking place today disrupts convention and tradition, while it counteracts the hegemony of the institutions and elites who try to manage and manipulate the cultural economy. The "passive audience" for media and the culture industries, if it ever existed in any totalizing way, has certainly disappeared.

Memes are signs, not hammers. They replicate ideas, not people, and even then only provisionally. When memes cluster together, the cultural meanings they generate become more numerous, intricate, and open-ended, not less so. Social imitation becomes exponentially more complex over time because the modes and codes of human communication have grown so elaborate. Memeplexes function as indeterminate discursive spaces—semiotic zones of social negotiation that are differentially represented and variably interpreted by human beings. The trends that characterize any historical moment, especially nowadays, invariably reflect a precarious and evolving cultural order.

All forms of cultural transmission have one thing in common: communication activity operating under selection pressure.[41] Like biological mutations, the cultural variations that social imitation produces are selected competitively. The introduction of some new information disrupts the cultural status quo while other information serves to maintain traditions and protect the institutions and individuals that benefit from them. Established cultural traditions, customs, habits, and preferences tend to be selected more often than unfamiliar interventions for the short term. To be out of step with our social groups threatens individual viability. But these regressive tendencies can be overcome. Any notion of cultures as shapers of individuals is out of sync with the new global realities. Rather than thinking of memes as powerful cultural viruses whose shaping influence cannot be avoided, therefore, we must reverse the logic: *ultimately, it is the memes that cannot escape the fate of being transformed.*

MAKING CULTURAL HISTORY

We're not the only primates to form cultures and transmit cultural information to other members of the species. Chimpanzees vary regionally in many respects, including what they eat, how they make and use tools, how they groom themselves, how they communicate, and how they hunt.[42] Among other culturally differentiated behaviors, some orangutans learn from others how to crack open nuts with stones, while other nearby groups do not.[43] Gorillas acquire more than forty skills, including communication signaling, tool making, and ways of seeking comfort that demarcate one group from another, depending on environment.[44]

Distributed human groups organize and adapt through behavior that takes place in elaborate networks of social communication. Over time, repeated interaction creates patterns of daily life that are known in common and with others taken for granted. The predictable behavior that results helps individuals develop sufficient common ground to feel safe and secure. Along the way a culture's defining qualities—the values and behaviors that separate "us" from "them"—strengthen to the point where they are not easily challenged, especially when access to outside cultural information is limited. Routine patterns of social interaction become cultural norms that reinforce the dominant cultural ideology and the structure of authority that goes along with it. Individuals and groups that benefit most from the entrenched traditions and existing institutions typically try to suppress change, especially when religious beliefs and practices are being asserted or defended. Rule breakers are identified and punished. Given the hegemony of cultural ideology and social practice, why should we expect cultures to change?

In the biological world, natural selection allows diverse organisms to adapt advantageously to their changing environments. Imperfect replication is best at spawning solutions because it offers the possibility of gradual but systemic change from previously evolved forms. This amounts to what British geneticist Steve Jones calls Darwin's great idea: life is a series of successful mistakes.[45] A seemingly unfortunate mutation of a presently flourishing type crosses the threshold to become an evolutionary success. An apparent loss or distortion of genetic information brings about a triumphant biological remodeling. Such genetic alterations can lead to surprising outcomes over long spans of time. Primeval amphibians stepped out of the soup and walked on land. Some mammals returned to spend their entire lives in water. Some fish can fly, and many birds never leave the ground.

The mutations that emerge cannot be too extreme, however, or the organism will not survive. The escalating growth of random variations would be fatal if biological mutations were not subject to the curbing effects of natural selection. Effective defense mechanisms are copied in nature. Consider the consequences that excessive distancing from the current model would bring to the viceroy butterfly (*Limenitis archippus*) and the species it imitates. The brilliant viceroy butterfly has nearly identical coloring, size, and shape of the more common monarch butterfly (*Danaus plexippus*), the taste of which repels predators. If significant alterations in either the monarch or viceroy butterfly were to occur, the changes would kill off the beautiful orange and black insects that rely on no other means of defense.[46] Mexico's Pueblan milk snake, which looks identical to the region's venomous coral snake, is among many other species that also benefit from physical mimicry. Even *Dracula* orchids in the cloud forests of Ecuador have evolved to resemble native mushrooms in order to attract mushroom flies, thereby increasing their chances of pollination.

Cultures do not create individuals. Individuals make up culture. In much the same way that random variation in genetic transmission leads to widespread biological modification that enhances the fitness of individual organisms, personal initiative can also bring about guided variation in cultural change.[47] The value of any cultural idea will finally be judged by individuals looking out for their personal interests. Like biological change, cultural development proceeds incrementally as a bottom-up, branching-out process that begins simply but produces outcomes of increasing complexity and greater functional value. In their book *Not by Genes Alone*, Peter Richerson and Robert Boyd describe the parallel between culture and biology nicely with a classic example: "Humans can add one innovation after another to a tradition until the results resemble organs of extreme perfection, like the eye."[48]

Genes can be pinpointed in stretches of DNA, and cultures can be seen in artifacts and art, but neither genes nor cultures are perceptible entities. They belong to another realm—an organizational frame, an a priori. Culture behaves "as if" it exists.[49] It appears to be organized, unified, restrained, and functional, but at the same time it is chaotic, disjointed, expansive, and dysfunctional. It is amorphous, too; culture is real in terms of lived experience and ideal in that people never conform perfectly to a cultural standard. Immutability and transformability, permanence and change are hard at work in every evolutionary process.

Cultural elements are subject to change from the moment they first appear. Information cannot jump from one brain to another without being exposed to a host of contaminating influences that open up possibilities for the production of cultural permutations. Innovation driven by individuals wrestles with the status quo driven by tradition and ideology. Like the spontaneous excursions of improvisational jazz, cultural nuances appear and attract their followers. Even Susan Blackmore, the British psychologist who argues that memes dominate cultural life, self-consciously presents herself to the world as a clear alternative to conventional culture. Her bouquet of red, blue, and green hair says anything but "I'm just another helpless victim of meme-driven mainstream culture."[50]

The biological principle that diversification of structure sustains the greatest amount of life also applies to the processes of cultural development.[51] Shared values and long-term familiarity may bring security and stability, but they tend to dampen potential creative sparks.[52] The more massive and urbanized a culture becomes, the more diverse and open it will be, bringing about more rapid change. Population growth allows more mutations to appear, making it possible for an increasing number of good solutions to be selected. Those solutions are most often generated by individuals operating in settings that reward free thinking and innovation. That's why big cities and online communities foster so much creative production. Communicative interaction that takes place in densely populated areas is more concentrated, efficient, and convenient, contributing to the relatively fast pace and improved quality of cultural change.[53]

The opposite is also true: change flourishes less quickly in more traditional settings, especially where education is undervalued and religion is prominent. Lifestyles in rural areas change less rapidly than in urban settings. Ethnic tribes and religious radicals reflect a mentality that requires absolute devotion and obedience. Destructive acts can then be carried out and defended by asserting cultural ideology. The signature act of many Islamic terrorists is to slit the throat of the enemy—directly attacking the organ that gives humans our greatest advantage over other species and, ironically, our best hope for minimizing the ideological and cultural differences among us.

Like all large social organizations, cultures tend to persist in their present form in part simply because they are so big and seem to have always been there. Yet cultures invariably change and usually improve. Like adaptations to existing biological forms, cultural traditions and practices erode and reconfigure in light of changing environmental conditions and opportunities. Each single alteration is modest, of course, just as any individual gene substitution within a complex organic adaptation is minor.[54] But over time and overall, societies worldwide have gotten better. Much better. How else do we explain the development of diverse and sophisticated human civilizations from their common primitive roots?

CULTURE REMIX

Historically, culture refers to human communities that occupy the same geographic territory, rely on resources close at hand, speak the same language, worship the same deities, and behave similarly across the spectrum of everyday activities. Traditional culture is composed of relatively stable biological, material, social, and spiritual forms that surround us from birth. It is a particular way of life shared by a community and shaped by characteristic values, rituals, beliefs, territory, and material objects.[55] Culture is inherently social; it makes available the frames through which we can know ourselves and others. Culture is "our way of doing things" and reveals "who we are" and "who we are not."

Human cultures nonetheless should not unambiguously be considered as places or conditions within which people conform or against which they are bound to struggle. The dynamics of cultural change are far more complex than that, especially now. Throughout nature, the organism, community, and context interact and evolve together. Environments tentatively represent what organisms do individually and collectively. Geneticist Richard Lewontin describes the integrated nature of biological evolution with the case of a common insect: "In fruit flies, which live on yeast, the worm like immature stages of the fly tunnel into the rotting fruit, creating more surface in which yeast can grow so that, up to a point, the more larvae, the greater amount of food available. Fruit flies are not only consumers, but also farmers."[56] The fruit fly thus creates a niche that fosters its survival. Cultural adaptation—change—takes place much the same way. For social species, the niches where biological change is constructed are composed of the totality of communicative interactions that are performed within a group. For humans, the niche of cultural development is the constant exchange of messages between people across the range of the unmediated and mediated social networks they occupy.

Cultural niches fluctuate wildly now. Fundamental oppositions that have long defined essential sociocultural dichotomies—self/other, individual/collective, insider/outsider, producer/consumer, and so on—are fading fast. Cultural experience has morphed from "read only" to "read/write."[57] Common ground that is created by the constraints of time and space and managed by cultural authority is less easy to identify. Cultural assumptions are being questioned like never before. In particular, media audiences and culture consumers have become more active, engaged, and discriminating than ever before.[58]

As communications technologies have evolved, the self-assembling, self-governing individual has climbed onto center stage in the theater of cultural experience. Conventional culture imposed by authority is giving way to personalized culture on demand. People meld traditional cultural elements like religion, language, food, and music with resources arriving from afar to create their fast-expanding cultural experiences and identities. The multicultural self engages the world as a creative project under constant revision. Much like programmers of radio and television outlets who organize the content of their channels to fit particular formats, individual persons can draw from an expanding field of resources to create cultural experiences ranging from the conduct of routine everyday activities to the shaping of the most profound aspects of cultural identity. Like traditional media programmers, individuals create, import, edit, store, and transmit cultural content. Many measure the response to their cultural production by counting "hits," "likes," and "friends"—the equivalent of commercial ratings success.

Personal cultural programming—where individuals mix and match bits and pieces of diverse cultural resources to satisfy their particular needs and interests and produce complex symbolic forms themselves—are becoming the dominant modes of cultural experience. The product of this cultural work is the individual's "superculture"—a constantly evolving matrix of symbolic forms and material resources that are especially relevant to the person.[59]

Cultural opportunities never distribute equally. Many people still lack basic education and health services and have little or no access to advanced communications technology or the ability to travel. But people with sufficient freedom, financial resources, and access to technology can exercise much greater control over their cultural experiences than ever before. Industry feeds the trend. From the first Kodak® still camera, record player, audiocassette recorder, and portable video camera to the Walkman®, MP3 player, personal computer, camera phone, and on to today's smart communication devices, consumer technology has become increasingly personalized and mobile. People can program their MP3 players with their favorite songs; create customized radio stations online from services like Pandora, Last.fm, Turntable, Slacker, or Ocarina; record their favorite television programs to play back when convenient; download and read e-books or watch movies on-demand; find restaurants and clubs on the go; check websites and communicate with friends anytime and anywhere.

But modern culture is as much about personal production as it is about consumption. Individuals use new communication technologies to gratify their expressive needs. Everyday people have become cultural authors and artists. Even a cultural statement as simple as posting a profile picture on Facebook® becomes a creative and highly personalized communicative act. People routinely employ the expressive technologies available to them to sample, fuse, style, pierce, and feel their way through their inner thoughts and feelings. Expressive behavior like this is fundamental to psychological health.

When Marshall McLuhan traced the history of communications development from oral to print to electronic, he showed that technological change is continuous and, like biological reversion, contains past morphologies and memories in its current structures.[60] The interplay between technological form and content produces what McLuhan called "hybrid energy" in much the same way that "hybrid vigor"—the production of superior variations arising from the crossbreeding of genetically different plants or animals—shapes the evolution of biological organisms. Radio and television refreshed the primordial roots of human communication by emphasizing orality, spontaneity, popular appeal, emotion, storytelling, and ritual.[61] Today's digital media stimulate involvement, expression, and community building, and mimic the rituals and social practices of the past while focusing on what's happening in the here and now. Each new media platform draws technically and discursively from previous stages. The Internet absorbed satellite and cable television, which had absorbed broadcast television, which had absorbed radio and film, which had absorbed print media, which had absorbed oral communication. Along the way, the silicon chip replaced the transistor. Digital superseded analog.

Just as nature shapes the evolutionary trajectory of the biological organisms that survive, successful new ideas are generated by people who are surrounded by scientific and technical knowledge and who work in environments that support innovative thinking.[62] Noise and error are common, even necessary.[63] Functional innovation drives the evolution of cultural complexity. The institutions where technical work takes place today must be flexible and adaptable. Constant give-and-take between divergent and convergent thinking drives the process. Apple® Computer may be the best example of this atmosphere in the field of personal communications devices. Open-source websites and software depend on input from below. Social networking sites monitor the behavior of their users and those of other sites to create new features. Musicians sidestep record companies to release songs online and interact directly with fans. Not-for-profit media organization WikiLeaks and hackers from individuals to highly organized groups like Anonymous dig out sensitive documents from deep inside government and private archives. The remarkable uprisings in the Middle East, North Africa, and south Asia, as well as the worldwide Occupy movement of 2011 and 2012, were all instigated by individuals and social networks that sprang up outside the established hierarchies of power. Changes in the technological environment stimulate increased variation in social and cultural behavior. Those new behaviors then alter the overall political and cultural landscape in a continuous spiral of reflexivity and adaptation.

CULTURAL DEVELOPMENT

By Charles Darwin's time, the true origin and natural history of life on the earth had been purposefully obfuscated for hundreds, even thousands, of years. Less than two centuries before Darwin was born, the Catholic Church had condemned Italian astronomer and philosopher Galileo Galilei as a heretic when he, using the emerging tools of physical science, validated the Copernican hypothesis: the earth revolves around the sun, not the other way around. The implications of an infinite universe containing other inhabited worlds were becoming shockingly clear: the cosmos was not made just for humans.[64] For his efforts, Galileo was forced to recant the truth of his scientific breakthrough and live out his last years under house arrest by order of the Roman Inquisition. Despite the profound disagreement with the Vatican, Galileo maintained a formal allegiance to the Church until he died, though he likely rejected Christianity personally. A hundred years later, Isaac Newton demonstrated how the laws of gravity and physical motion confirmed Galileo's insight and provided irrefutable evidence of the underlying orderliness of the universe. For his part, Newton remained a believer in an early version of intelligent design.

Galileo's and Newton's daring work surely helped inspire Darwin to pursue lines of scientific reasoning that he knew would meet with deep-rooted resistance. Yet the basic facts were clear, and Darwin felt compelled to put forward his understanding of how species come into being. In *On the Origin of Species*, Darwin described a world brimming with biological and cultural diversity, where mutation and selection produce constantly evolving, better-adapted animals and plants in nature's wild play of forces.

Qualifying the notion that the drive for sheer individual survival will prevail at any cost, Darwin saw how intentional, morally guided behavior can alter evolutionary outcomes. He remained steadfastly optimistic even as he pondered the destructive conflicts that have developed historically between human groups. Nonetheless, the human capacity for critical reflection, he thought, ultimately makes it possible to rebel against our self-preserving tendencies and the influence of dominant culture so that our natural "social instincts and sympathies" can eventually extend to "the men of all nations and races."[65] This would be a long-term project. Cultural differences within and between nation-states certainly have not gone away since Darwin wrote so positively about the human potential. But, happily, we do continue to demonstrate the ability to "rise above our origins" and to "overthrow the tyranny of natural selection," as Richard

Dawkins observes, in order to plan "gentler societies," including the widespread presence of social welfare, progressive taxation, and the secular codification of ethical behavior.[66]

In less than fifty thousand years—a mere blink of the eye in evolutionary terms—humans have progressed from exchanging simple signals in the hunt for large animals to connecting instantly across the planet for any number of reasons. Spoken language, body decoration, iconography, and writing have all accelerated cultural growth tremendously. As the world's symbolic species, we have gradually become less directly dependent on natural resources and more reliant on information and each other.[67] Tremendously useful new channels for creative expression have appeared. Cultural boundaries are breaking down. Even the defining geopolitical development of the current era—globalization—has progressed from what was mainly a transnational economic system to a much more complex and democratic cultural phenomenon powered by information and communication.

Darwin recognized how communication contributes to cultural development by comparing the ways information moves around in different societies. The language of Darwin's day may offend, but the point he made in the nineteenth century remains correct today: "In semi-civilized countries, with little free communication, the spreading of knowledge will be a slow process."[68] The constructive potential of unfettered communication rests at the heart of human development. More than 150 years after *On the Origin of Species* appeared, the United Nations Development Programme uses only slightly different phrasing to advocate the same idea for building healthy multicultural democracies. According to the United Nations, tolerance and appreciation for differences within countries and open access to cultural resources arriving from outside national borders are necessary for making social progress in any nation-state, especially the more tradition-bound parts of the developing world.[69]

Biological evolution has produced an unfathomable array of lifeforms over the past three and a half billion years. In but a tiny sliver of that time, cultural development has ushered in science, technology, democracy, civil institutions, even little pills that prevent pregnancy and prolong sexual performance. Nature is pure information; it provides us with endless resources. Culture is applied knowledge; we can improve what nature gives us. The trajectories of our natural history reveal a meaningful difference in the way change takes place and what it means for our collective future. Evolution is about variation in nature; development is about innovation in culture.

CHAPTER 5

COMMUNICATING MORALITY

After finishing a long-term project describing and categorizing barnacles, an investigation that further established his credentials as a scientist, Charles Darwin anxiously resumed labor on a draft of what he would call his "abominable volume"—*On the Origin of Species*. For Darwin, the late 1850s were difficult times. His youngest child, Charles Waring Darwin, died suddenly of scarlet fever—the second of his ten children to be taken by illness. Grief disrupted Darwin's life in the English countryside.

That wasn't the end of it. On Friday morning, June 18, 1858, just ten days before the death of his son, Darwin received a small packet sent from Ternate, then a Dutch colony in the East Indies. Another naturalist—considered by experts in London to be more a specimen collector than established scholar, and with whom Charles Darwin was only slightly acquainted—asked Darwin to pass along a theoretical paper to one of his closest friends, geologist Charles Lyell. Darwin took the opportunity to read the paper himself. In what has become familiar lore, Darwin was flabbergasted. The document, written by Alfred Russell Wallace, spelled out in very similar terms the same central idea that drove Darwin's own work in progress. Alone, fighting fever in the jungles of New Guinea, and working independently from Darwin, Wallace had also formulated the principle of natural selection. The news overwhelmed Darwin.

Luckily, however, Darwin's life work pondering the basis of evolution and what he called the "transmutation of species" was not scrapped. Charles Lyell and botanist J. D. Hooker did what they could to assure Darwin that his share in the breakthrough explanation of natural selection would be preserved. They arranged a session of the prestigious Linnean Society in London where Wallace's and Darwin's papers were presented jointly to the scientific public. Encouraged by this exposure but tormented by the loss of a child and stunned at having apparently been upstaged by Wallace, Darwin restlessly finished his scrupulous manuscript. His publisher, John Murray, began to take orders for *On the Origin of Species* on November 22, 1859.

Charles Darwin himself tried to drum up a favorable response to his controversial book. He mailed off copies of the book with flattering letters to many of the most distinguished members of the Victorian scientific establishment. Nevertheless, four days before the book went on sale, the London magazine *Athenaeum* set the tone for the hostile reaction Darwin's book was about to receive—an animosity toward the work that persists even more viciously in some quarters today. Pointing to what he considered to be *The Origins*' utter disregard for moral values, an anonymous reviewer asked in a sarcastic and accusative tone: "If a monkey has become a man—what may not a man become?"[1]

The reviewer insisted that Darwin's book should be repudiated because of its damaging theological implications. Other fits of moralistic outrage followed. Adam Sedgwick—Darwin's old teacher and professor of geology—wrote him to say the book caused him more pain than pleasure. Sedgwick charged that evolutionary theory threatened to plunge humankind into moral turpitude, ethical degradation, and rampant agnosticism. The controversy culminated in a public debate at the British Association for the Advancement of Science at Oxford the following year. Without Darwin present, Bishop Samuel Wilberforce of the Anglican Church and zoologist Thomas H. Huxley of the Royal School of Mines clashed raucously over who could speak with greater authority on the origin of life: theologians or scientists. The debate clearly mapped the contrasts between scientific inquiry and religion over moral values. *On the Origin of Species* had become the dividing line.

The conflict over the ownership of morality as a personal virtue continues even today, sometimes in most unpleasant ways, as the scandalous case of a famous California soldier dramatically reveals. Young Pat Tillman—an avowed agnostic—gave up an extremely lucrative career as a professional football player to join the United States Army in patriotic response to the terrorist attack in New York in 2001. Months later, the celebrated, handsome soldier was shot to death in the mountainous terrain of Afghanistan under highly suspicious conditions. United States military authorities conducted an investigation to determine the cause of Tillman's death. They claimed he was cut down by enemy fire. Tillman was posthumously awarded the Silver Star for valor.

The Tillman family was not satisfied with how the investigation had been carried out or with the conclusions reached. Disputing the military's explanation, unofficial sources told the family that Tillman might have been killed by friendly fire, perhaps even by "fragging"—the intentional killing of a fellow soldier. The Tillman family went public to expose what they believed was a compromised, possibly corrupt inquiry into Pat's death. The officer who led the official investigation responded on national television to the Tillman family's angry refusal to accept the military's findings. The officer said their reaction could be explained only by a lack of belief in God: "These people have a hard time letting it go. . . . If you're an atheist and you don't believe in anything, if you die, what is there to go to? Nothing. You are worm dirt."[2] The officer reasoned that Tillman's parents considered the death of their son to be spiritually incomprehensible because, for them, death was the end of everything; they didn't believe Pat ascended to heaven. The family's lack of faith, the officer thought, barred them from experiencing proper consolation. Sadly, Pat Tillman, the nonbeliever, could only become "worm dirt."[3]

Morality marks the difference between right and wrong in human affairs, a distinction that seemingly has always been confounded with belief in God. In his day, Darwin worried that heated controversies about the nature of human origins and the question of God's existence would provoke emotional reactions that might prejudice even some educated readers against all his ideas. To avoid that outcome, he took great care not to excise God from the book. He even mentioned "the Creation" and "the breath of the Creator" explicitly in subsequent editions of *On the Origin of Species*, a blatant concession he later came to regret.

Charles Darwin's careful phrasing could not conceal the most obvious and inevitable implication of evolutionary theory: natural selection made the existence of God a useless hypothesis. By summoning a most impressive body of empirical data, he was able to show that morality emerges from nature, not from divine intervention, and that selection processes continue to regulate moral relationships among living organisms in the biological world.[4] Moral behavior, he argued, derives from the give-and-take that occurs naturally among members of a species as they work out differences between individual and group interests.

Moral reasoning by humans depends on the ability of a person to empathetically imagine the role of another person in a system that requires interdependence for mutual survival. The Golden Rule—which appears in some form in every cultural group—originated as a biological, not theological, principle. The Bible and other religious texts reflect lessons people everywhere have learned from nature. Many lower animals also have a version of the "do unto others" principle. In fact, when compared to other species, humans don't always emerge as the superior moral beings; some lower-order groups act with consistent moral character while human populations do not. The realities made public by Darwin on this matter presented serious problems for religious authorities. If morality springs upward from the shadowy origins of biological life and doesn't alight from divine heights, then religion stands to lose its mighty grip over human affairs. The challenge to conventional thinking that was posed by evolutionary theory threatened to disrupt a dominant cultural and political ideology supported by religious authority.

Natural theology—a worldview held by some nineteenth-century intellectuals that did not depend on religious texts but still claimed God created the universe—influenced Darwin's pre-evolutionary thinking. But natural theology was also rendered irrelevant by the theory of natural selection. By the time he finished writing *On the Origin of Species*, Darwin was no longer willing to grant creationism any scientific purchase whatsoever. He wrote, "It is easy to hide our ignorance under such expressions as the 'plan of creation,' 'unity of design,' etc., and to think that we give an explanation when we only restate a fact."[5]

Scientific research conducted since then confirms the fact that high-flying morals are not the ethical property of religions, nor are they cultural extensions of theology. Moral systems and the rule of law draw historically from religion, but the moral principles expressed by religious groups have an indigenous origin. Moral values arise and evolve because reciprocity and community play crucial roles in nature. As the diverse contemporary work of Marc Hauser, Frans de Waal, Marco Iacoboni, Michael Tomasello, and many others has clearly shown, any claim that moral behavior originates in culture and religion has the causal ordering backward.[6] Two indisputable facts support the claims about moral behavior advanced by Darwin: (1) the basic psychological mechanisms that underlie moral behavior in humans operate in other living creatures, too, especially in other primates, and (2) the roots of human morality predate the establishment of religion.

Some critics of Darwin's work feared that giving any scientific credence to evolutionary theory would undermine traditional Western standards of morality. But the danger to moral standards was not at issue then and isn't now. Human beings don't have to be religious to act with moral integrity. Our foremost moral commitment is to ourselves and others, not to a belief system. Religion can help people organize their desire to be virtuous, but many people who naturally want to be good and do good deeds actually become corrupted by religion's influence. By granting moral consideration to some people but not to others, even within particular tribes and faiths, religion compartmentalizes morality and infuses it with the potential for discrimination and destruction. Women and homosexuals have been especially victimized by religious moralizing. Centuries-long standoffs between Sunni and Shia Muslims and between Catholic and Protestant Christians reveal how moral principles can disintegrate in the face of religion's influence, even within relatively homogenous groups.

The moral imperatives of the Torah, the Bible, the Koran, and Islamic hadiths seem on the surface to be universally relevant and constructive, but they were never meant to apply to everybody. The sanctity of moral actions encoded within major religious texts pertains only to members of the various in-groups or candidates for conversion. "'Love thy neighbor' didn't mean what we now think it means," writes Richard Dawkins. "It meant only 'Love another Jew.'"[7] When New Testament Gospels (Matthew 5:44; Luke 6:27) refer to loving one's enemies, what is literally said is *diligite inimicos vestros*. Matthew and Luke beseeched believers to love their *private* enemies (*inimicos*)—the ones who formed part of one's religious community but allowed them to be free—and encouraged believers to despise their *public* enemies (*hostis*).[8] The Muslim equivalent of the Golden Rule appears in a hadith: "None of you truly believes until he wishes for his brother what he wishes for himself."[9] But the "brothers" to whom the rule applies are fellow Muslims. Confucius had a reverse spin on the same idea: "What you do not want done to yourself, do not do to others."[10] There is nothing special or divine about the principle of moral reciprocity. It's mundane biology. To call it the word of God and then divide humanity into groups of insiders and outsiders who sometimes even kill each other only reflects another principle of nature: the destructive potential of tribal loyalty.

DEFANGING NATURAL SELECTION

On the Origin of Species made clear the discomfiting fact that nature's constantly evolving magnificence results in large measure from outright competition and acts of ruthless destruction. Natural selection appears to spread nothing but selfishness, exploitation, and indifference to the suffering and weakness of others. Devoid of any trace of compassion, natural selection seems unable to provide the moral foundation for any lasting system of ethical values. For theists, this raises a troubling, if rarely acknowledged, question. How could a compassionate God concerned with projecting moral values possibly have created living creatures that routinely cannibalize one another? For evolutionists, the issue is less puzzling. Nature is absolutely contingent; the fate of the natural world must be open and indefinite. Ruthless destruction is simply part of what goes on.

Darwin's Victorian opponents had a fair point when they perceived the immense difficulty in reconciling the existence of a moral world with the picture of life that is presented in *On the Origin of Species*. It's difficult to imagine how any set of uplifting moral concepts could possibly anchor a world propagated by natural selection. Self-preservation and reproductive potential rule all of life. Other people exist mainly as vehicles for spreading one's genes. Natural selection leads individual organisms to do whatever is necessary to ensure reproductive success—cheat, manipulate, beguile, murder, rape, even deliver sincere and true messages. All of this should be carried out with absolute impunity so long as the goal is met. The "is" of now differs greatly from the "ought" of any future moment and cannot, therefore, shape moral behavior positively. Natural selection, it seems, cannot favor self-sacrifice and restraint. Darwin himself was acutely aware of this problem: how can morality exist when natural selection yields so obviously to the immense pressures of the present?

Thinking this way, a world driven by the pressure of natural selection seems depressingly grim. Close observation of natural phenomena, however, reveals a much more complex and hopeful picture. Like any good scientist, Darwin always addressed possible exceptions, qualifications, and additions to his evolving theory of natural selection. The emphasis he placed on complexity and doubt together with his extraordinary skills of observation led him to a brilliant insight. He noticed that selfless behavior routinely goes on inside the hives, mounds, and nests of bees, ants, termites, and other insects. Their actions seemed to contradict selection's rude realities. Why, for instance, do bees go on living in hives that are split not just into classes of males and females but also into sterile (neuter) and reproductive castes? Deprived of reproductive capacity, neuters act with selfless devotion to their sisters so that when the right time comes, an unrelated female will become the new queen and provide the colony with offspring. Neuters are completely marginalized in the process, resigned to caring for offspring that can never be their own. This seems to be an indisputable instance of self-denial. Such behavior could not occur according to the strictures of natural selection, Darwin thought, assuming as he did that selection qualifies as a universal biological law.

What, then, makes up the true costs and benefits of altruistic behavior in the animal world? How can the principles of natural selection possibly lead to self-sacrifice? Is it simply because the totality of the hive, mound, or nest is greater than the sum of its individual insects? Do hives, mounds, and nests operate as magnified single entities—superorganisms that follow the principles of natural selection at a more abstract group level? Perhaps, but the better explanation is that *the homes of these insects function as integrated systems held together by pervasive communication activity wherein each behavior is ultimately motivated by self-interest.* Unaware of genetics, Darwin had no inkling that the insects he studied were taking care of their own interests by acting according to instincts that indirectly perpetuated their genes.

Ever since Darwin's early attempt to explain altruism in *On the Origin of Species*, evolutionists have come up with three basic solutions for understanding why and how selflessness exists within the competitive confines of nature. The first theory posits that altruism can confer an overall advantage on a group. Individual sacrifice is justified in the name of beneficial social coordination and collective action. This is *group selection.* But in nature, altruistic advantages within a group are initially conferred on siblings and other relatives of individuals. In this way *kin selection*, the second solution, puts the individual solidly back into the equation of evolutionary advantage while also spreading beneficial effects over a general population. A third kind of unselfish behavior occurs when unrelated individual organisms perform services for each other and expect some kind of return on the investment. This is the familiar "I'll do something for you and then you'll do something for me" formulation. Or, "I'll be altruistic for you if you'll return the favor." This arrangement represents the fundamental principle of *reciprocal altruism*. Each of these solutions helps explain how altruism works in the social life of animals and humans.[11] We shall now discuss them one by one.

GROUP SELECTION

Although Charles Darwin never used the term "group selection" in his books, he wrote extensively about the roles of morality and community in natural selection, including intergroup competition among humans: "It must not be forgotten that . . . advancement in the standard of morality will certainly give an immense advantage to one tribe over another. There can be no doubt that a tribe including many members who, from possessing in a high degree the spirit of patriotism, fidelity, obedience, courage, and sympathy, were always ready to aid one another, and to sacrifice themselves for the common good, would be victorious over most other tribes; and this would be natural selection."[12] Organisms of many species behave morally within their communities because there is an evolutionary payoff for individuals that belong to the groups that prevail. Animal populations that exhibit pro-social traits are more likely to survive and reproduce than those that don't.[13] They develop a "group mind" that favors cohesion, cooperation, and individual sacrifice. Social insect colonies have become the standard examples of superorganisms that exhibit this kind of group behavior. As evolutionary biologist David Sloan Wilson puts it, "Just as the mind of an individual is not contained within any single neuron or hormone, a social insect colony has a mind that is not contained within any single insect."[14]

Individuals act in concert with other individuals to promote their own interests in the context of the group. The apparently self-sacrificing instincts that are evident in ant and bird populations, for instance, evolved from repeated interactions taking place among interdependent, proximate individuals.[15] Altruistic alliances that develop among members minimize the group's internal contradictions and nurture social stability and security. Groups that act collectively develop a kind of moral intelligence that forms a platform for evolutionary success. Group selection tones down selfishness to the point where the aggressive impetus of natural selection becomes displaced by the advantages of collective altruism.

Cooperative group behavior thrives because it will be selected by the vast majority of individuals in the group. But crucial questions remain. How can the abstract idea of the greater good of the group be enforced in populations whose members also dedicate themselves to individual self-preservation and reproductive survival? How can altruistic behavior be advantageous for organisms when the beneficiaries are not even their own offspring, as is the case with sterile insects that die by the thousands without passing on their own genes directly? And what does group behavior tell us about the morality of human populations?

The whole of the group develops over time from the totality of ongoing relations established among individuals. The gathering together of individual organisms in natural environments often creates an impression of group uniformity. It suggests the mounting of a collective force that is stronger than its individual parts. A school of fish, a flock of birds, a cloud of mosquitoes, a herd of African antelopes, or even a human crowd seen from a distance appear this way. It is tempting but misleading to assume that these and other groups act jointly because they have surrendered themselves to the disinterested pursuit of moral uprightness and collective harmony. What appears to be a coordinating force is often an illusion of orderliness and consensus.

The better explanation is that each individual's chances for survival and reproduction are increased by forming groups and cooperating. Most individual organisms stand scant chances of surviving when left alone. Animals living in open spaces—the plains, the skies above, unprotected waters—form groups to make themselves and their progeny less vulnerable to predators. Grouping forces predators to have to expend more energy searching for prey. A large group is more difficult to find than lone individuals or small groups traveling across a widely distributed area. Individuals also move as a group for protection at the moment of attack. The danger zone between predator and prey should be as large as possible.[16] Tight clustering pays off; individuals situated in the middle of the pack enjoy the best prospects for survival.

Despite the costly investment of great energy, animals keep moving to survive. Internal communication drives the locomotive behavior. An individual initiates the action. Each organism then adjusts its movements by responding to the actions of other individuals, directing and redirecting the shape and movement of the crowd. Messages sent by behavior produce impromptu changes. The order, organization, and structure of the group's internal interaction emerge phenomenologically from repeated behavior. Still, the groups often move unpredictably, which makes them less vulnerable to predators. Group movement is produced spontaneously by the totality of signals transmitted inside the cluster. There is no architect, no choreographer, and no conductor of the flow.[17] Coordination emerges through what appears to be fast-changing "group think," but the patterns of movement reflect only the sum of individual behaviors in flux.[18] The behaviors of the individual and of the group are being created simultaneously by the totality of individual acts performed within the group.

From support for vaccines and special education programs in schools to labor union strikes, military field tactics, and calls for flash mobs to assemble, adaptive group behavior extends into extremely complex social situations among human populations. Even groups of our closest cousins, troops of chimpanzees, don't organize their communities or execute their hunting expeditions in very complex ways. They help others proactively but draw the line quickly on how much they will invest in others' well-being. Chimps and the other primates lack the communication skills and shared sense of intentionality that humans possess.[19] They communicate through physical actions, gestures, facial expressions, and simple vocalizations that express emotion more than transfer information. Their social behavior becomes instinctual over time and lacks the flexibility, creativity, and immediate adaptability that human communication encourages.

In *Darwin's Cathedral*, David Sloan Wilson explains the power of group selection by focusing on the unique role organized religion plays in human civilizations. Wilson argues that religion persists as a powerful social institution because religious communities adapt very successfully as groups.[20] Religious groups function like beehives. They "adapt their members to their local environment enabling them to achieve by collective action what they cannot achieve alone or even in the absence of religion."[21] The actual appeal of religion thus has mainly to do with evolutionary advantages offered by community. Moral codes professed by religious groups are integrated into their systems of belief. The belief system helps buttress the functionality of the group, providing a tangible benefit to individual members of the congregations. Wilson refers to the functionality of religion as its "practical realism."

Human cultures have codified and institutionalized empathy, cooperation, and integrity in a wide variety of grandiose discourses, narratives, rules, and legal documents. Morality is well recognized and regulated by human groups. Yet even the most sophisticated democratic societies don't always operate with moral clarity. For example, public discussions and elections supposedly reflect individual participation in forming the collective will of a democratically constituted population. But after paying lip service to honorable goals designed to advance the common good, elected representatives frequently act in ways that promote their selfish interests and those of their cronies. To maintain public trust, democratic societies have had to develop elaborate systems of checks and balances and thousands of rules and regulations to avoid sliding into mob rule or being crushed by demagogues. For the same reasons, religious fundamentalists often prefer to obey their faith-based moral codes and laws rather than follow those that are created and enforced by secular authority.

Altruism is not just about sacrifice and conformity for virtuous reasons or for the autonomous advantage of the group and its individual members. The benefits of altruistic behavior and moral development can also improve the group's capacity for annihilating other groups—tribes fighting tribes, as Darwin said—and not just for reasons of self-defense. Combativeness, violence, and warfare often escalate in the name of ideology and the irrational self-righteousness that collective behavior can inspire. Osama bin Laden and George W. Bush both justified their support for murder in the name of social justice. At the group level, mayhem is the dark side of altruism.

Levels of natural selection do not exclude or compete with each other. Geneticist George R. Price established mathematically that natural selection can act simultaneously and progressively at the individual and group level.[22] The best way to think about the role of group selection in broader perspective is to consider it "multilevel selection."[23] Selection acts first at the individual level. If something works against individual interest, it can't have a positive effect for the group. On the other hand, if mutational changes prove to be advantageous to the individual, they can also benefit the population as a whole and are more likely to be adopted.

KIN SELECTION

The motives for altruistic acts performed in the natural world do not reflect unequivocal support for the higher interests of the group. Degree of genetic relatedness between individuals also plays a role. The cost of an altruistic act becomes justified when it benefits the genes of the altruist. Living organisms protect the offspring with whom they are genetically related. By helping relatives survive, the altruist accomplishes something almost as good as propagating one's own genes. The closer the genetic connection, the stronger the protective bond. This perspective was described in the 1960s by William Hamilton as "inclusive fitness."[24] It is explained clearly by John Maynard Smith and Eörs Szathmáry in *The Origins of Life*: "When contemplating the spread of a gene causing its bearer to behave in an altruistic manner, one must consider not only its (adverse) effect on the bearer, but also its (beneficial) effect on the bearer's relatives, multiplied by the respective degree of genetic relatedness."[25]

Charles Darwin called this phenomenon the principle of "family selection."[26] To better understand how kinship relationships affect population genetics at the human level, consider the explanation put forward by evolutionary biologist J. B. S. Haldane:

> Let us suppose that you carry a rare gene which affects your behavior so that you jump into a flooded river and save a child, but you have one chance in ten of being drowned, while I do not possess the gene, and stand on the bank and watch the child drown. If the child is your own child or your brother or sister's, there is an even chance that the child will also have this gene, so five genes will be saved in children for one lost in an adult. If you save a grandchild or nephew the advantage is only two and a half to one. If you save a first cousin, the effect is very slight. If you try to save your cousin once removed, the population is more likely to lose this valuable gene than to gain it.[27]

These calculations are complicated and not made consciously. Instincts take over. Genes reveal that even the most communitarian insects are not completely selfless creatures. They act altruistically to maximize their overall genetic success while foregoing the benefits of individual sexual reproduction. Again, it's a matter of probabilities. For instance, because the reproduction rate in a beehive is exceedingly low—in the range of one in tens of thousands—it makes sense for female workers to assist the queen, with whom the workers are genetically related, in her reproductive efforts. In exchange, the queen gives the female workers food without which they can't survive.

Considering the grand scheme of sexual reproduction, however, this trade-off may seem unfair. Worker bees give up their individual long-term reproductive potential—the key to biological immortality—only for immediate survival. But from the viewpoint of the individual bee, it pays to keep the queen's genes viable because all the workers are genetically related to her. The worker can't have its own colony, but it can help the queen generate more workers, thereby helping to keep the genetic line alive. A selfish foundation underlies the bee's unselfish behavior.

Selfishness rooted in kinship relationships is not the only motivator of animal altruism. Environmental circumstances can also trigger acts of apparent self-denying cooperation. For example, some varieties of the African starling—bright blue and orange birds that live in the unpredictable climate of semiarid savannah grasslands—delay their own sexual reproduction to help raise the progeny of others, helping to ensure that more young birds will survive.[28] The fact that the birds they help may or may not be genetically related to them—a characteristic of cooperative behavior that is present in many other species—has led some scientists to argue that any strong influence of genetic relatedness on altruism may be limited only to certain populations.[29]

While genes aren't directly visible to life-forms, information about genetic relations within a population can be extracted from communicated messages. Many lower and higher organisms recognize their kin directly through the presence of corporeal marks or a common odor.[30] Some birds identify relatives by their feathering or other visible characteristics. Evening Grosbeaks, for example, discern subtle variations in the white bands on their wings and in the thickness of their beaks. Particular patches of feathers on the sides, back, and head of the Yellow-rumped Warbler serve as reliable indicators of kinship. American coot females kick out hatchlings that don't physically resemble their firstborn, suspecting that the egg was placed in the nest by a freeloader. Even some plant species recognize and favor their relatives by processing information about light patterns and chemicals that are released into the ground and air.[31]

The chemical-material foundation of scent often functions as a decisive form of sensory communication that is given off and received by virtually all living organisms, even many lower-order species and varieties. The scent of various butterfly species, for example, identifies family members while attracting mates and excluding predators. Ants' long and sensitive paired antennae are used to smell food and kin. In order for a queen bee to maintain her rule over the hive, she must produce a sufficient amount of pungent pheromones—a chemical that triggers a behavioral response in another member of the same species so that bees entering hives in which no relatives live are killed.[32] Among other species of fish, North American bluegill live in nests composed of mixed paternity where they sort out and protect kin by scent.[33]

Favoring relatives over nonrelatives is also universal behavior among humans.[34] Males carefully examine the physical features of their offspring to confirm fatherhood, but this remains an ambiguous indicator of genetic relatedness. To increase chances that the offspring of any sexual pairing carry the rightful male's genes, human societies formulate various kinds of legal agreements—marriage and divorce being the most common—to help regulate sexual reproduction. The social contracts often carry severe penalties for disobedience. More draconian methods have also been used, and some are still employed. Chastity belts, female genital mutilation, female covering, restrictions on women's mobility and professional development, close family monitoring, swearing sexual fidelity before God, stoning adulterers, and many other techniques have been devised to better assure males that the children they care for (when they do) are biologically their own.[35]

The distribution of genetic relatedness that underlies the principle of inclusive fitness applies throughout the natural world. What birds, bees, wasps, and ants do with their kin is perfectly consistent with what human fathers and mothers do with their children. The power of genetic relatedness shows up in everyday behavior. People tend to display more kindness and generosity toward their own offspring than toward the children of their siblings. That's because humans share half their genes with their offspring, compared to their nephews and nieces, who carry only half of that half.[36] Parenting is inherently sacrificial. Sorting out genetic relationships can bring heavy emotional consequences. A parent who loses custody of a child also loses part of himself or herself. That's one reason why custody disputes generate such hefty revenues for divorce lawyers. Lie detectors, DNA paternity tests, and other kinds of genetic examinations used to prove family relationships have even become common audience-grabbing ploys used on reality TV and talk shows.

RECIPROCAL ALTRUISM

Why smile at a passing stranger who will never cross your path again or leave a tip for a waiter in a city to which you'll never return? Why obey the traffic signal when no police officer is present or give your seat to a pregnant woman on a crowded bus? Why send monetary donations to victims of war or natural disaster on the other side of the world? On the opposite side of the spectrum, why do family members often take each other for granted, frequently despise one another, and sometimes even hate their siblings to the point of murder—as in the archetypal case of Cain and Abel—while acting kindly toward unrelated strangers?

Group selection and genetic relatedness explain much about cooperative social behavior because animal groups benefit from their associations with other members of their species and organize themselves within kinship systems. But group membership and genetic connectedness can't be the only means for cultivating altruism. Unrelated organisms also have to create cooperative relationships for species to develop in ways that benefit their populations. The success of a species ultimately depends on the quality of relationships that develop among unknowns.

Mutual benefit that emerges among organisms irrespective of their proxemic or genetic commonalities is what evolutionary biologist Robert Trivers calls reciprocal altruism.[37] The logic that guides reciprocal altruism is simple and familiar: a favor is provided and something is given in return. For humans, the expectation of receiving something exists at many levels. A person might do someone a favor today, for instance, and expect something back quite soon. In other cases, no immediate payoff is expected. Yet self-interest never leaves the equation. Even when an individual does something just "to make the world a better place," that action implies that conditions will improve for the individual's descendants ("For the sake of our children, and our children's children . . ."). For religious people, the return on investment might take the form of increased chances of getting into heaven. Religious ideology always presumes moral authority. It's not insanity that drives religious terrorism, as geneticist Spencer Wells puts it, "rather it is the God-given certainty that what one is doing is morally just."[38]

We've come to expect altruistic behavior from our fellow human beings, but other mammals, including wild dogs, chimpanzees, hyenas, dwarf mongooses, and naked mole rats also behave according to the principle of reciprocal altruism in some situations. Muskrats expose their bodies to predators to distract them and to warn other muskrats of a common enemy. Babbler birds compete fiercely among themselves to be recognized as superior and exclusive altruists.[39] The case of vampire bats is particularly illustrative. Bats forage every night and return to the cave. Some are successful in the hunt, some are not. A threatening situation continually exists because bats that do not nourish themselves with blood begin to weaken significantly in less than three days. Whether related or not, bats that roost together donate blood to those who need it. The favor is commonly returned. Loafers and cheaters are recognized and expelled from the community.[40]

The benefits of reciprocal altruism don't pertain just to intraspecies interactions. In Richard Dawkins's estimation, "Reciprocal altruism works because of asymmetries in need and in capacities to meet them. That's why it works especially well between different species; the asymmetries are greater."[41] Take, for example, the cleaner fish (*Labroides dimidiatus*). This species of very small fish enters the mouths of much bigger fish, running the risk of being eaten, to clean edible residue off the large fish's teeth. In return, the small fish swims safely into the large fish's mouth when it needs to avoid predators. On the Galápagos Islands, small lizards scamper unmolested across the broad backs of sea lions to eat bothersome flies, and finches peck away at harmful parasites on the hides of the enormous tortoises.[42] The biosphere of the crowded and inhospitable Keeling Islands' coral reef in the Indian Ocean described by Darwin in his first monograph thrives because of collaboration and communication among species.[43] In *Where Good Ideas Come From*, Steven Johnson explains that "what makes the reef so inventive is not the struggle between the organisms but the way they have learned to collaborate—the coral and zooxanthellae (algae) and the parrotfish borrowing and reinventing each other's work."[44] Interspecies trust like this does not develop overnight. Cooperation evolves only after countless interactions build up counter-instinctual habits that themselves later become instinctual.

Survival is a cooperative project at every biological and social level. Popular science writer Carl Zimmer points out that "even the cells in the human body cooperate. Rather than reproducing as fast as it can, each cell respects the needs of the body, helping to form the heart, the lung, and other vital organs."[45] In a similar way, each of our genes needs a network of interdependent replicators to survive. Like cells and genes, individual human beings, even those who share no immediate genetic heritage, must connect and work cooperatively to succeed. Members of healthy societies take care of each other and benefit individually from the effort.[46] Gross social inequality tears at the human psyche, creating anxiety, distrust, and mental and physical ailments.[47] It takes a village, as the popular saying goes.

Social cooperation benefits individual villagers, not just some general notion of the group. That's why a primary motivation for demonstrating moral excellence is to be positively reviewed by a potential mate as a candidate for sexual selection. Expressions of sensitivity, empathy, kindness, and cooperation put personal qualities on display, draw positive attention, and create opportunities for sexual reproduction. This principle applies broadly in the animal world and is communicated from one individual to another. For example, aggressive male chimpanzees share meat, papayas, oranges, cassava, and other food with females of reproductive age. The males' willingness to raid farms to gather food attracts female attention while at the same time intimidating competing males.[48] The male house finch, a small bird common to western North America, presents potential mating partners with choice bits of food. The female accepts his proposal by mimicking the actions of a hungry chick—a particularly revealing check of his fitness, because male house finches feed their mates during breeding and incubation and later feed the offspring. Acorn woodpeckers gather nuts for other family members when food is abundant or scarce, thereby promoting their genetic interests while demonstrating their individual virtues.[49] Sentries among Arabian babblers, highly social birds indigenous to the Middle East, refuse to eat when food is offered them during their periods of vigilance over the flock. Babblers also try to help when a fellow babbler is caught up in a net, and adult males receive food even when they don't forage properly. Performing these altruistic acts marks individual birds as fit for genetic reproduction.[50]

Reliable communication conveys altruistic intent among related and unrelated individuals. Even the insects that Charles Darwin observed to be altruistic were the ones that had developed the most sophisticated communication systems. Honest, true, and reliable communication allows all living organisms, especially humans, to strike an efficient balance between the immediate selfish interests of the individual and the long-term success of the group. Group success encourages individuals to expand their altruistic behavior. Owing its origin to a common ancestor, reciprocal altruism is the default form of human sociability in all cultures.[51] The implications for human development are clear. As Darwin wrote, "Each man could soon learn from experience that if he aided a fellow-man, he could commonly receive aid in return."[52] What Darwin meant by "learning from experience" is the favorable outcome of communications activity. Cooperative communication transforms the selfish tendencies of natural selection by encouraging reciprocal altruism. Communicating responsible social behavior is valued because it demonstrates an investment of energy—an evolutionary cost for the individual—for advancing the common good. Altruism can thus best be understood as the communication of personal qualities through behavior.

COMMUNICATING TRUST

Cooperative communication and moral behavior originated because to help others was to help oneself. Ironically, selfishness gives rise to altruistic individuals.[53] But outright selfishness never rules the day. Restraints on behavior are required for social living and are favored by natural and sexual selection for their survival value. Even if evolution were to have produced populations composed almost entirely of utterly selfless organisms, those unspoiled groups would live under constant threat of invasion by a few selfish individuals. The selfish ones would assault the meek, reproduce themselves, and eventually transform a community of self-sacrificing organisms into more selfish beings. Rules for restraining self-interest eventually evolved, as groups found ways to rein in, punish, and expel rogue individuals. Selfishness can win out for some individuals inside a group for the short term, but, as we've seen, characteristically altruistic groups outlast the selfish ones.[54]

Variations on the principle of mutual benefit—reciprocal altruism, symbiosis, mutualism—run deep in the natural world. The social nature of evolution drives everything. Cooperation assumes that each participating individual understands that the needs of other individuals must be taken into account. Yet some degree of self-interest never leaves the equation. To be viable for the long term, cooperation must bestow benefits on all parties. As John Maynard Smith and Eörs Szathmáry point out, "Two co-operating individuals must do better than they would if each acted on its own."[55] As one consequence of this survival strategy, cooperative relationships create divisions of labor in animals and humans.

Cooperation is required for cultural and moral development. We cooperate instinctually, but our hominid ancestors' skills as rudimentary communicators had to have emerged prior to the development of complex forms of social cooperation. Repeated cooperative behavior then crystallized into habits that spurred the further acquisition of communication skills, the additional expansion of cooperation, and the deepening of moral sensibilities in general. All social communication requires recognition of the other and can inspire moral behavior: the *sender* of a *message* uses *channels* to find a *receiver* or *receivers* who respond or reciprocate (*feedback*).

Human cooperative communication evolved through three main stages: *requesting* (I want you to do something for me), *informing* (I want you to know something because I think it will help or interest you), and *sharing* (I want you to feel something so that we can share attitudes and feelings together).[56] The last two stages mark a crucial difference between humans and the other primates. Advancing to the third stage establishes the common ground on which cultures develop their moral values and ethics. Our cooperative nature and flexible modes of interaction give us substantial control over our individual and collective lives.

The impact of repeated social behavior on genetic structure seems to have hardwired humans to be empathetic and cooperative.[57] Mirror neurons stimulate individuals to feel the pain and joy of others.[58] People want to be good; it's to their advantage. For this reason, common moral precepts developed in all societies: do unto others, don't kill, don't steal, avoid adultery and incest, and take care of children and the weak.[59] The most powerful social institutions everywhere formed around a shared sense of morality and exist most effectively when their members agree to abide by basic rules of ethical behavior. Despite the similarities in moral inheritance, however, we don't all act the same way in practice. Moral instincts have been channeled into particular modes, codes, and rules of conduct that reflect and shape local cultures, much like language and music have taken on varying characteristics in different settings. So, for example, most Japanese give high priority to community (*uchi*) over self, strict Hindus and Orthodox Jews believe they can achieve purity through dietary restrictions, and pious Muslims respect moral authority by refusing to insult or depict "the Prophet."[60]

Throughout modern history, the impetus for collaboration has been driven by trade and barter practices where interpersonal trust was essential. Maritime commerce among the diverse populations surrounding the Mediterranean Sea in the fifth century set the stage for economic globalization.[61] European capitalism later diminished monarchical and feudal authority, further spurring creation of social networks among trading partners that crossed social class and cultural lines.[62] Interpersonal trust has become more difficult to grant, as economic systems have expanded greatly since then. Still, tangible benefits continue to accrue from a willingness to have confidence in others. Even in today's risky global economy, business organizations that trust their own employees and trading partners usually fare better than those that habitually harbor suspicion and distrust.[63]

Personal integrity and pro-social behavior generate positive evolutionary outcomes for individuals and groups. Yet deceptive communication—lies, tricks, and outright exploitation—remains common in nature even among members of the same species.[64] Deception is evolutionarily useful for getting food or sex, for example. For that reason, deceptive behavior is built into our social conduct, even before language. Small children will cry insincerely to manipulate adults. Believing our own lies can serve to strengthen our self-confidence as individuals and as groups. But there is a price to pay.

Being deceived means the loss of invested time and effort—an exorbitant cost in evolutionary terms—so most deceivers will ultimately be discovered and punished by their confederates.[65]

Positive, conscious moral decision making can overcome the self-interested tit-for-tat that underlies all forms of altruism. The advanced state of moral development among humans marks the greatest distinction between humans and animals, according to Charles Darwin, who defined a moral person as someone "who is capable of reflecting on his past actions and their motives—of approving of some and disapproving of others."[66] Our moral sense tells us what's right and wrong in the cultural contexts we inhabit; our conscience reproves us if we disobey. As Darwin argued: "Ultimately our moral sense or conscience becomes a highly complex sentiment—originating in the social instincts, largely guided by the approbation of our fellow men, rules of reason, self-interest, and in later times by deep religious feelings, and confirmed by instruction and habit."[67]

The "deep religious feelings" of which Darwin spoke became the primary cultural channels through which moral behavior was institutionalized in human societies. But moral codes also created a sense of self-righteousness among religious groups and provoked bloody conflicts between them. Although group stability is always subject to adaptive remodeling, religious authority rarely welcomes change. Completely absent from other animal groups, organized religion arrived much later than moral behavior in the evolution of our species and, ironically, for reasons we explain in the next chapter, often inhibits the spread of ethical principles.

CHAPTER 6

COMMUNICATING RELIGION

In a chapter on the evolution of mental powers in *The Descent of Man*, Charles Darwin tells a humorous story about his dog:

> My dog, a full-grown and very sensible animal, was lying on the lawn during a hot and still day; but at a little distance a slight breeze occasionally moved an open parasol, which would have been wholly disregarded by the dog, had any one stood near it. As it was, every time that the parasol slightly moved, the dog growled fiercely and barked. He must, I think, have reasoned to himself in a rapid and unconscious manner, the movement without any apparent cause indicated the presence of some strange living agent, and that no stranger had a right to be on his territory.[1]

Dog owners recognize this behavior. Parents might see it when their children jump away from automatically flushing toilets. A powerful and threatening hand must be responsible for things we cannot easily explain.

Darwin told the dog story to show how believing in spiritual agencies—like those that magically move parasols or flush toilets—can "easily pass into the belief in the existence of one or more gods" and eventually into religious devotion.[2] Despite religion's bloody legacy, unswerving belief in the existence of an omnipotent God in Darwin's day was considered to be ennobling. Darwin himself didn't completely give up on Christianity until he was forty.[3] For the vast majority of the world's population today, belief in God—the one true God—continues to dignify individuals morally and socially. This is true even in some modern societies, especially in the United States, where more than 60 percent of the population still believes that acting morally requires belief in God.[4]

Philosopher A. C. Grayling argues that, because of the deference granted religion, "the religious voice in the public square around the world has been vastly overemphasized and vastly potentiated" even in secular states.[5] For instance, England still formally recognizes an established church, and in that country, tax revenues support faith-based schools, religious programs appear regularly on public broadcasting services, Anglican bishops loom in the House of Lords, and prayers open sessions of Parliament. Daily collective worship in public school is required by English law. With the Tory government's emphasis on local control over curricula and other matters, evolution may be taught less in the classroom in the future, and the fast-increasing number of faith-based schools may be permitted to introduce creationism into science courses.

Imbued with a powerful sense of national exceptionalism that shaped their country's identity around religious freedom and devotion from the very beginning, Americans embrace religious faith much more deeply than does the majority population in the United Kingdom and Europe. Many Americans tend to conflate religious commitment with patriotism. Religious ideology was used to justify both sides in the American Civil War. Christianity was promoted as the antidote to the evils of communism in the realignment of political and military power following World War II. The United States became one nation, *under God*, with liberty and justice for all when an amendment was added to the Pledge of Allegiance in 1954. Less than a decade later, Reverend Martin Luther King's lofty Christian rhetoric inspired the mid-twentieth-century civil rights movement. Television evangelists like Pat Robertson, Jerry Falwell, Jimmy Swaggart, and T. D. Jakes began filling the airwaves with nationalist politics and cultural conservatism toward the end of the twentieth century.

The twenty-first century has brought other kinds of religious influence. By far, America's most-watched cable news channel, Fox News, proudly builds its image as the defender of "God, country, and family."[6] The Creation Museum in Petersburg, Kentucky, draws hundreds of thousands of visitors annually with its story of young earth creationism, during which time humans and dinosaurs magically roamed the earth together. The Texas Board of Education requires that social studies textbooks play up the false idea that the nation's federal government was founded on Christian articles of faith and that they play down the Jeffersonian principle of separation of church and state. Teachers in many American public schools are encouraged to treat evolutionary theory skeptically and balance it with "creation science." Private religious schools teach creationism any way they choose.

Some 40 percent of Americans still believe humans were created by God within the last ten thousand years.[7] Even more think that Jesus Christ will return to the earth by 2050.[8] Predictions of a definite Rapture date circulate widely every few years. About 83 percent of the American public believes God answers their personal prayers.[9] A larger number are not completely sure their prayers will be answered, but they pray anyway, especially in time of need.[10] The federal government promotes an annual National Day of Prayer.

Of the 535 members of the 112th Congress who took office in 2011, only six did not claim a religious affiliation, and none said they were atheist or agnostic.[11] Two years earlier, the moral barometer of suitability for the American presidency had been measured largely by how convincingly the candidates could express personal commitment to religious faith. Republican candidate Mike Huckabee, an ordained Baptist minister, denied evolution and promised he would be "a Christian leader." Mitt Romney, a Mormon, told a national TV audience that "freedom requires religion," and that European societies have become "too busy or too 'enlightened' to venture inside [cathedrals] and kneel in prayer." Hillary Clinton's campaign website claimed her Methodist religion is the "driving force in the way she lives her life, including voting on issues."

And Barack Hussein Obama—who had been labeled a Muslim by some right-wing extremists—had to assure the nation over and over again that he "prays to Jesus every night." No politician can be successful on Chicago's South Side—Obama's home territory—unless he networks effectively in the area's historical African American megachurches. But to Obama's great credit and to the dismay of traditionalists, his inaugural address in 2009 included the line "We are a nation of Christians and Muslims, Jews and Hindus, and non-believers." Later calling for a renewed commitment to science education, Obama pointed out that Abraham Lincoln's birthday—the occasion for a speech a month after he took office—was shared by another great historical figure—Charles Darwin. And although Obama asks God to bless the United States of America at the end of every speech, a concession that all American presidents feel they must make regardless of their true feelings about religion, his critics pointed out that the president didn't attend church during his first eleven weeks in office.

The Republican field of candidates in the 2012 presidential campaign was filled to the brim with religious zealots. One of the top early candidates, Governor Rick Perry of Texas, issued a proclamation for "Days of Prayer for Rain" when drought hit the state the year before. He also preached to thirty thousand evangelicals gathered in Houston, asking for God's help in fixing America's economic decline. Congresswoman Michele Bachmann insisted that earthquakes and hurricanes were messages from God to fix our culture, and her husband's rehabilitation center was devoted to ridding gay people of their sacrilegious malady. Newt Gingrich made a campaign issue of his conversion to Catholicism as a sign of redemption for his past "moral failures." Rick Santorum's entire candidacy was rooted in religious qualifications, and he claimed that Barack Obama was not fit to be president because his policies were not "based on the Bible." Mitt Romney never flinched from his Mormonism, and Jon Huntsman, another Mormon, tried desperately to create a more moderate image but never let go of his faith. Facing a tough bid for reelection, Obama kept praising the Lord everywhere he spoke.

The politics of religion isn't just a game played by politicians. Plenty of American entertainers promote religion too, often to great financial benefit. Popular singer and activist Lady Gaga mentions Jesus and uses religious imagery in much of her music, especially in the emblematic "Born This Way," her gay liberation–themed album. "Jesus is my virtue," she sings, "and Judas is the demon I cling to." Gaga succinctly captures the contrast: religion is a defense against the baser instincts of the self. But it's much more than that. Religion is a multipurpose resource for redemption, meaning, and immortality. Faith gives the poor a strong identity and hope for the future. In crime-ridden areas, religious belief serves as an alternative to the perils of the street. For anyone who is weary, worried, or weak, religion can easily become a psychological comfort zone, social network, and cultural space. Individuals who choose to be religious deserve to be respected as human beings. But the commitment to religious faith brings consequences beyond personal safety, comfort, and hope that must also be addressed.

WHO'S MORAL?

Surrendering oneself to God and accepting the tenets of religious belief are still considered by many to be the noble and humbling acts Darwin described as the "sacred virtue" of abject submission to a tribe.[12] Abrahamic scripture declares that a moral order was imposed on humanity by God.[13] Religious believers engage with a community of like-minded people who, by the mere pretense of their apparent subjugation, assume a lofty moral stature. The stakes are high; winners and losers emerge in a zero-sum game of who gets to heaven and who doesn't. Apostates, converts, skeptics, and the followers of other faiths suffer the ultimate punishment.

The youngest of the Abrahamic religions—Islam—is particularly forceful in its claim of moral righteousness and sense of submission. As Immanuel Wallerstein explains in *The Decline of American*

Power, "The Christians claimed that they had fulfilled the Jewish law and therefore supplanted it with a new and final revelation. The Muslims in turn claimed that they had built on the wisdom that they had inherited from the Jews and Christians with a new and truly final form of commitment to Allah."[14] The commitment is absolute. The Arabic term *islam* means "to submit." A *Muslim* is one who submits.

Activist Ayaan Hirsi Ali tells of how she felt as an Islamic teenager in Somalia wearing a black body-covering hijab: "It sent out a message of superiority: I was the one true Muslim. All those other little girls with their little white headscarves were children, hypocrites."[15] Many Muslim men show their depth of faith by growing their beards long. One sign of personal commitment to Islam is a dark callus on the center of the forehead that comes from repeatedly banging one's head on the floor during prayer. Sharia law legitimizes and extends intolerance and prejudice. Even today, huge majorities of Muslims in Egypt, Pakistan, and Jordan favor stoning people who commit adultery, whipping or cutting off the hands of robbers, and imposing the death penalty on individuals who leave the religion. Muslim populations in more modern and ethnically diverse nations like Turkey, Lebanon, and Indonesia are far less likely to endorse these measures.[16]

It wasn't Christianity that appealed to Hirsi Ali when she immigrated to Holland as a young woman. It was the order, cleanliness, kindness, and freedom of secular European cultures. That freedom relegates religion to personal choice and gives it a much less dominant role in culture. It provides far greater opportunities for women. Apparently, relative godlessness does not diminish a culture's moral fiber, at least not in the Netherlands, which ranks high among the most humane and prosperous countries on the earth and is one of the least religious. Norway, Denmark, Iceland, Sweden, Canada, Belgium, Switzerland, Japan, and the United Kingdom also rate high on quality of life and low on religiosity.[17]

By contrast, the world's least culturally and economically developed countries today are the most religious.[18] The deleterious effect of strong religious influence stems largely from two of religion's most central and damaging historical characteristics: freezing women out of more diverse and productive social roles and clamping down on innovation. These two acts are interrelated. Gender equality—which remains conspicuously absent in deeply religious cultures—has proven to be the most reliable indicator of how strongly a society supports the principles of tolerance and egalitarianism. Tolerance and egalitarianism feed creativity, dissent, risk taking, and technological integration—the foundations of innovation. Innovation positively influences a society's economic development.[19]

ROOTS OF THE PROBLEM

How did religion's inviolable belief systems and unassailable declarations of moral righteousness—the "terrible certainties of faith,"[20] to use Richard Dawkins's memorable phrase—come into being and propagate themselves so strongly and widely? Charles Darwin reasoned that religion must serve evolutionary purposes because religions appear everywhere and have so much in common.[21] He noted that even the physical characteristics of spiritual worship, including "the uplifted hands and turning up of the eyes in prayer," appear in all religions.[22] Supernatural belief emerged over the millennia as one of the human universals.[23] Magical thinking, superstition, and clan behavior all predate the religions we know today. Yet the origin and exact adaptive value of "so many absurd religious beliefs . . . that have become, in all quarters of the world, so deeply impressed on the mind of men" was unclear to Charles Darwin.[24]

Even today it's not certain whether the commonalities that exist among religions spring from shared cognitive structures like those that gave rise to language and morality, or if they derive from ideas and practices that have spread randomly by means of genetic or cultural drift.[25] Regardless of its origins, religion's development was not evolutionarily inevitable; we humans could have organized our cultures around other ideas, and in some cases we did. In particular, the elements of nature on earth, the sun, the moon, and the constellations were worshipped. Still, a combination of unique qualities goes a long way to explain religion's emergence and sustainability. The word *religion* derives from the Latin verb *religare*, which means "to tie or bind." The major religions bind people together psychologically and socially by giving them two mighty safeguards: personal protection on earth and the promise of everlasting life in heaven.

Communication facilitates both ambitions. Humans communicate interpersonally to create belief systems and form religious communities on the earth. Everyday life becomes permeated with religious practice to the point, in some cases, that no distinction is made between religion and culture. Consent becomes implicit as the faithful purchase a safe place for themselves by socializing and conversing, sometimes exclusively, with other believers. They immerse themselves in singing, praying, reading scripture, making pilgrimages, celebrating holidays, and participating in countless other customs, rites, and ceremonies. Vocalization and body movement excite and reinforce religious programming. Nonbelievers become unavoidably subjected to the constant presence of religious thought, too, and sometimes unwittingly participate in its construction, even when uttering expressions like "Goddamnit!" "What the Hell?" or "Bless you!"

Second, many religious people believe they communicate with God directly or through his emissaries when they pray for guidance, consolation, or favors. The "imaginary friend" relationship is constructed in a way that extends normative social interaction. British zoologist Robert A. Hinde describes how this occurs: "Many people have a need to share their experiences with others, and may gain comfort from talking over their experiences with a supposed other being. . . . In addressing a deity it is almost impossible to believe that a response will not be forthcoming, just as when one speaks to a friend one is sure he will reply."[26]

Communities provide the physical safety, sense of emotional belongingness, and social stability that are essential for the well-being of all primates.[27] In a classic study, sociologist Émile Durkheim described how the communal aspects of religion—especially social solidarity and close familiarity with others—form the very foundation of human societies.[28] This is the "practical realism" of religion described by David Sloan Wilson. Religious beliefs function like genetic mutations: "They arise arbitrarily and only the ones that work are retained by imitation and selection."[29] Successful belief systems are "user friendly," Wilson says, because "they reduce the complexity *of* the real world to motivate a suite of behaviors that are adaptive *in* the real world."[30] Religious behavior promotes many evolutionarily advantageous values and regulates society accordingly.[31] Individuals are required to conform to social rules especially as they concern marriage, reproduction, and child raising. Religion facilitates political movements and creates the interpersonal trust needed to engage in commerce and to wage war.

But practical realism alone cannot adequately explain religion's broad appeal. Religion also makes the idea of God understandable, relevant, and familiar. Ruminations on the most profound existential questions naturally arose as humans' expanding cognitive capacity made conscious reflection and reasoning possible. The first attempts to probe the origin and meaning of life were exercises in rational thinking. But without the benefit of scientific knowledge, the mysterious workings of nature could be explained only by inventing supernatural forces. Psychological and social commitments made to a higher power and to fellow believers helped individuals deal with the uncertainties of life and the certainty of death.

Religion is a triumphant mindset. Of course, any grandiose, nonscientific explanation of phenomena that exists so far from normal comprehension can only be hopelessly reductionist and misleading. Still, for religion to have become the powerful ideological and cultural force that it is, individuals must be willing to entertain magical explanations of experience on equal footing with the credibility of every other kind of information.[32] The human neurological condition makes this possible. While broad regions of the brain react somewhat differently when individuals are asked to compare everyday facts, the brain makes no discernible distinctions among content sources when processing information.[33] The brain responds essentially the same way to ideas that are big or small, data-based or belief-based, reasonable or unreasonable.

The nineteenth-century American philosopher William James defined religion as "the feelings, acts, and experiences of men in their solitude, so far as they apprehend themselves to stand in relation to whatever they consider as the divine."[34] Driven by the search for meaning, community, and immortality fueled by the limitless power of the imagination, emotional realism forms the psychological foundation of religious belief. As science journalist Nicholas Wade points out, the emotional commitment to a group "must be so fierce and transcendent that men would quite readily sacrifice their lives in its defense."[35] Religious behavior became the solution to this emotional requirement and, in Wade's view, developed gradually into an instinct.[36] Tribes later became demarcated by their religious beliefs, which were passed along from generation to generation. Religion is not the only social institution that can serve humankind's basic practical and emotional needs, but it does so with incomparable effect across a wide range of cultural groups.

By the time our ancestors left Africa, all the basic elements of religious behavior were in place.[37] The particular features of religious thought and conduct, however, developed much later. Charles Darwin wrote that even the "idea of a universal and beneficent Creator does not seem to arrive in the mind of man until he has been elevated by long-continuing culture."[38] Instead, he said, a general belief in "unseen or spiritual agencies" first arose among cultural groups and formed the psychological templates that would facilitate religious development.[39] The mind was being conditioned to regard superstition and magical thinking as normal. God became an inferred agent—the creator of unexplainable, uncontrollable, and threatening events, and later, a source of personal security and salvation. Speculation about every miraculous feat—creating the earth, deciding who will enjoy an afterlife or suffer endless torture, answering prayer requests, sending out guardian angels, and so on—could then be attributed to God. The list of things God intends, does, requires, and forbids, as well as wild conjecture about the consequences of his judgments, attest to the enormous power of the human imagination. Creating the perception that heaven and hell are actual physical locations, or that religious outliers will be severely punished on a real judgment day, is a remarkable invention of the restless mind.

Surviving one's death promises the maximum payoff for the believer, and the dream of immortality is taken seriously by many people. Why wouldn't it be? As Robert A. Hinde argues, "Mortality runs counter to biological propensities conducive to survival." It is "difficult to imagine a state of non-being . . . one cannot imagine a world without one's presence, because one is inevitably present behind the eye that is seeing that world."[40] Moreover, the glorious reward for a lifetime of righteous behavior (and penance when gone astray) appears to be within reach because no irrefutable evidence can be gathered to disprove it. Of course, no evidence supports the idea of immortality either, but that's not required. Apparently, the inability to access God materially only adds to the attractiveness of the idea that he exists and that immortality is possible. Indeed, the founders of young religions like Mormonism and Scientology have had to make their claims increasingly eccentric to attract a sizeable following in the crowded theological marketplace.

SPIRITUAL COMMUNICATION

The social structure of early human groups changed when our ancestors formed tribes of hunters and gatherers and settled into progressively stable communities. An emerging division of labor gave rise to the role of specialists. The tribal shaman—an earthly agent thought to be capable of summoning good and evil spirits and of giving symbolic presence to the spirits—and the artist—whose efforts represent the first pictorial communication—interacted in ways that profoundly influenced cultural development.

Shamans expressed the dreams, fears, fantasies, and relationships among life-forms that were depicted in primitive art. The images centered on humans' triumph over nature—a key theme that is preserved in religious discourse today. Snakes, birds, horses, bulls, stags, and elephants—each symbolizing a particular kind of power—were common subjects. Eerie supernatural worlds appeared together with hunters and their bounty. The earliest traces of religion's beginnings—cave paintings and artifacts produced by European early modern humans, formerly referred to as the Cro-Magnon—reflect distinct ceremonial qualities that reinforced and spread superstition and magical thinking.[41] The drawings were created in the caves' deep recesses—locations that lent themselves acoustically to drumming and chanting.[42] Music has driven religious behavior from the beginning. The idea of an omnipresent god or gods and a sweeping sense of the sacred were being formed by the developing forms of mediated communication.

Early humans planted intermediaries between themselves and nature: God and his flesh-and-blood emissaries on the earth. The shaman would become the rabbi, the priest, the minister, the mullah, the monk, the guru. Caves served as the first houses of worship, and chanting gave the spiritual rituals a sonic dimension. Natural psychotropic substances consumed by the shaman likely fueled the incantations.[43] Graves became more ornate, the dead were interred wearing symbolic beads and garments, and burial ceremonies grew ever more elaborate.[44] Ancestor worship emerged. Spiritual meditations, prayers, and other forms of worship became taken-for-granted cultural practices. Art, music, dance, and language evolved together in ways that shaped religious rituals from the beginning.[45]

Symbolic creativity played a key role in all early cultural development, including the expressive power of spiritualism and religion. The first cave art fused skeletal images of human beings with depictions of living creatures to create hybrid animals that appropriated the beasts' strength and endurance. The hybrids later grew more common and more fantastic. Mermaids originated in Assyrian mythology. The centaur and minotaur sprang from the creative imagination of the Greeks. Christians merged Jewish folklore with the legend of a deeply empathetic prophet to create a stirring narrative. Muslims added new stories, rituals, and a different kind of prophet to those inherited from Jews and Christians. Despite their diversity, some themes can be found in the symbolic production of nearly all mainstream religious groups. For example, the image of the human soul ascending triumphantly to heaven appears widely in many types of religious art and conveys the impression that, unlike animals, humans can achieve immortality.

The natural world has contributed a great deal of raw material to the development of religious iconography: the lion, the fish, the olive branch, and the lamb have all become important religious symbols. One symbolic form inspired by nature has a particularly interesting set of connotations. Originating in the Middle Ages, the Christian symbol for eternal life is the peacock. Decorative feathers from the splendid bird are still used today to symbolize the Resurrection during Easter services. The magnificent display of the male peacock's feathers later became the evolutionists' icon of sexual selection.

The link between faith and feathers is not just ironic. The lure of religious belief and practice interfuses deeply with gendered sexual selection. As Charles Darwin observed, throughout human history "religion and love have been strangely combined." He described how the sacred "holy kiss of love" corresponds in religious ritual to the kiss "a man bestows on a woman, or a woman on a man."[46] Spiritualism and the objectification of the female form appear together in the early cave art. From the start, religion has served men as a legitimizing channel for the control of women, for the propagation of males' genes, and as a way to deny outsiders access to "our" females. Churches, mosques, and synagogues function as pools of potential sex partners for certified candidates. Aren't nuns the brides of Christ, the everlasting husband? Muhammad had twelve wives. Joseph Smith Jr., founder of the Mormon Church, took between thirty and forty wives, and the founder of the religion's university, Brigham Young, upped the ante to fifty. Even Jesus of Nazareth enjoyed the close companionship of Mary Magdalene to the very end—a relationship that is not adequately explained in the Bible or by ecclesiastical authorities.

Many religious authorities are obsessed with sexual attraction. They fight carnal temptation by prohibiting premarital, extramarital, and homosexual sex; postmarital sex in some traditions; masturbation; and naughty thoughts generally. Communication between men and women in tradition-bound religious groups is restricted to such a degree that healthy human relationships are severely impaired. In some cases, males and females are kept apart in schools, in places of worship, even in all public spaces. It's no wonder that celibate suicide bombers are motivated by the idea of an endless sex party in the afterlife. As journalist and well-known "antitheist" Christopher Hitchens put it, the problem with the jihadists "is not so much that they desire virgins as they *are* virgins: their emotional and psychic growth irremediably stunted in the name of God, and the safety of many others menaced as a consequence of this alienation and deformation."[47]

BELIEF AND SCIENCE

Reflecting on his first trip by horseback to the stunning Yosemite Valley in Northern California's Sierra Nevada mountain range in 1868, naturalist John Muir wrote, "Everyone needs beauty as well as bread, places to play and to pray in where nature may heal and cheer and give strength to the body and soul alike."[48] But as he studied the crevices cut deep into the majestic granite prominences of the valley over the years, Muir also correctly reasoned that the jagged landscape must have been carved by glaciers thousands of years ago.

Muir took great spiritual and physical inspiration from nature. It was his place to play and pray in. Yet because he understood that the earth's surface was and continues to be shaped by natural forces acting over long stretches of time, John Muir was also making an argument for evolution. His informal theory of the geological history of Yosemite Valley mirrors the conclusions drawn by Charles Darwin when he explored the Chilean cordillera thirty-three years earlier.

John Muir was one of the world's first environmentalists. He founded the Sierra Club and convinced American presidents to establish a unique and extensive system of national parks. Spiritual and scientific thoughts were always copresent in Muir's writing. This is not an uncommon way of seeing things. In the minds of many individuals, including some accomplished scientists, religion and science do not have to cancel each other out. In a particularly famous case, Francis Collins, the former director of the National Human Genome Research Institute in the United States, committed to Christianity while admiring the wonders of nature in the wake of a parent's death.

Many ordinary people hold both evolution and religion in their heads with little or no discomfort. They may learn about evolution in school, for example, or they may visit a natural history museum where evolutionary facts are plainly described. Then, many of those same people attend synagogue, church, or mosque with no strong feeling of unrest. A remarkable example of this phenomenon appears in Richard Dawkins's British Broadcasting Corporation series *The Genius of Charles Darwin*. Weeks after Dawkins took a group of secondary school students on a field trip where they gathered fossil evidence and noted geological formations pointing to the actual age of the earth, many reported they found the experience interesting but still preferred to believe the creationist explanation.

Some people say that evolution probably acts on organisms that presently inhabit the earth, including humans, but they credit God for putting everything here in the first place.[49] Others hedge their bets on the species: lower life-forms may have evolved and continue to change even now, but humans do not. Still others downplay the literalness of the Book of Genesis and treat the religious volumes metaphorically, poetically, or selectively to accommodate scientific explanations while maintaining faith in God. Or perhaps it is God that is evolving. God has come to be regarded by populations across the globe as more universal, tolerant, and less violent over the years.[50] Today's kinder, gentler, more open-minded God contradicts the core principle of an unchanging and timeless deity but makes it less necessary to choose between religious faith and scientific fact.

The fact that religion delivers evolutionary benefits or can be cleverly reconciled with reality does not discredit Dawkins's view that religious belief in a world brimming with scientific knowledge isn't just irrational but is delusional and deceptive. Religious authority derives from documents that were written long before science replaced folk philosophy as the superior mode for explaining the natural world. At the time the Bible was written, people believed the earth was flat.[51] Many of the claims made in religious legends and books—like the story of the Israelites being freed from captivity in Egypt and led across the desert to the Promised Land—are demonstrably false.[52] It's not even clear which prophet—Jesus or Muhammad—was being referred to as the messenger of God in inscriptions on the Dome of the Rock in Jerusalem.[53] "Facts" have become extremely pliable through the ages because religious discourses were being shaped in ways that reflected competing interests in the political and cultural power struggles of various historical periods.

Philosopher and theologian Augustine of Hippo, a convert to Christianity who became one of the most important developers of the faith in the late fourth and early fifth centuries, quoting the Book of Isaiah, issued a famous dictum on knowledge: *credo ut intelligam* (I believe so that I can understand) and wrote an essay, *Of Faith on Things Unseen*.[54] The tradition expressed by Augustine is long-standing. God becomes the absolute truth and religious faith the governing framework for generating knowledge. Faith cannot be overruled. This is what former American presidential candidate Mike Huckabee meant when he said human beings are "not supposed to understand the Bible. If you do, your faith is too small." This is also what Father Timothy M. Dolan was saying on the occasion of his promotion to Catholic archbishop of New York when he implored his followers to "trust in what cannot be seen" and not to rely solely on information based on "empirical, scientific evidence." Dolan, now a cardinal, was following a biblical mandate. Proverbs 3:5 instructs the faithful to "trust in the LORD with all your heart, on your own intelligence rely not."

Science, on the other hand, produces theoretical explanations that are free-floating, transcultural, open to critique, and subject to change. Doubt may threaten religion, but it drives science. For this very reason, many nonbelievers feel uncomfortable labeling themselves as atheists. Just as religious believers cannot prove that God *does* exist, skeptics cannot prove that God *does not* exist. It's not completely out of the question that God might show up quite unexpectedly someday. Atheists can reduce the force of any latent reservations they may have about their atheism, however, by appropriating an analogy originated by philosopher Bertrand Russell early last century: it also cannot be proven that a teapot is not currently orbiting a planet somewhere in the vast reaches of the cosmos.[55] A responsible and rational person always keeps an open mind. That's why many nonbelievers claim to be agnostic in a scientific sense and atheist in a personal sense.

The origins of science predate the birth of modern mainstream religions. The Greek philosopher and mathematician Thales of Miletus—the first person to predict the date of a solar eclipse—argued as early as the fifth century BCE that nature follows natural laws. Documented visible movements of the planets revealed that they follow a demonstrable order; the movements were not a stream of unfathomable events orchestrated by mythological gods or heroes. Centuries later, physical scientists developed that base of knowledge to successfully expand the idea of a closed cosmos to that of an infinite universe not centered on the earth.[56]

The Greek philosophers established the clear benefits of knowledge derived from careful observation and well-reasoned interpretations of what is observed. Yet despite the benefit of this scientific advance, as Stephen Hawking points out, Christians later "rejected the idea the universe is governed by indifferent natural law. They also rejected the idea that humans do not hold a privileged place within that universe . . . a common theme was that the universe is God's dollhouse, and religion a far worthier study than the phenomena of nature."[57] The power of the religious apparatus over science is strong. The Catholic Church continued to hold sway over the development of all scientific thought in the West until the sixteenth century and still maintains widely publicized official positions on science, as do "authorities" from other religions.

Logic and reason often get shoved aside where matters of faith are concerned, even among individuals who otherwise act sensibly in their daily lives. Terminally ill religious people frequently make life-and-death decisions about medical treatment based on what they think God wants them to do rather than what their family doctor, spouse, children, or the scientific information recommend.[58] Intercessory prayer becomes a way to summon extra-personal help, akin to rubbing a rabbit's foot or hanging a dream catcher above a bed.[59] The magic of faith healing appeals to people far beyond the domain of religious fantasy. Americans spend billions of dollars every year on vitamins and alternative medicines with no credible scientific evidence proving their effectiveness. Millions of parents refuse to vaccinate their children because they still fear the vaccines cause severe health problems like autism, despite powerful scientific evidence that completely debunks that false belief.

God is a gloss, a covering expression, an intellectual cop-out that supposedly accounts for all that is knowable or popularly imagined. It's a dangerous idea. Without something against which to measure the validity of religious belief, emotional vulnerabilities can be exploited, and social conformity becomes easier for authorities to inculcate and manage. That's why totalitarians love belief. Systems of thought and structures of authority founded on fear and sentiment accentuate the uncomplicated, the extreme, the excluding, and are prone to bigotry and hate. This kind of thinking filters down to individual believers. Personal deficiencies are turned into superiorities: nonbelievers have a problem and we have the solution. The weight of such twisted logic degrades the quality of human relations "just as in genetics there is the danger that mutations of lethal recessive genes will emerge to cause crippling or fatal diseases" in a physical body.[60]

Answering scientific questions with religious explanations was even more common and bold in Charles Darwin's day. In *The Expression of Emotions in Man and Animals*, for instance, Darwin attempted to explain an intriguing phenomenon, blushing, which he called "the most peculiar and the most human of all expressions."[61] Many experts at the time believed blushing is God's way of instilling in people a signal of moral accountability by exposing their shame, guilt, or embarrassment. Sir Charles Bell, a Scottish anatomist and surgeon, argued that God designed blushing to show up only on the most exposed parts of the body—the face, neck, and upper chest—specifically for the purpose of revealing moral violations. This led Bell to conclude that blushing "is not acquired; it is from the beginning."[62] Thomas Henry Burgess, a medical doctor in London, claimed that God created blushing "in order that the soul might have sovereign power of displaying in the cheeks the various internal emotions of the moral feelings" revealing to everyone when "sacred rules" had been violated.[63]

These notional interpretations of blushing made no sense to Darwin, who countered with a scientific explanation: blushing is an inherited, involuntary, physical change in the movement of blood through the capillaries that results from positive or negative personal attention. Blushing can be caused by feelings of shame, Darwin acknowledged, but these sentiments spring from social and cultural conditioning that derives from evolutionary processes, not from displeasing God. The person who blushes shows that he or she cares about the perceptions of others, an indication of healthy social integration. Blushing can also be caused by shyness and modesty. The "blushing bride," for example, communicates purity, modesty, susceptibility to male attention, and desirability as a healthy sexual partner ready for genetic reproduction. Blushing is a selected trait.

Religious socialization today depends on the same kind of

magical thinking that shamans used to impress their fellow tribespeople or that some Victorian doctors employed to explain blushing. The obedient child brain is shaped by the forces of natural selection.[64] Because children rely instinctually on their parents to protect them, adult authority becomes associated with survival in the mind of trusting and gullible youngsters who, unlike the offspring of most other species, are unable to fend for themselves for years. Children pick up basic moral lessons imposed by adult instruction and come to believe that God, like their parents, watches them with a critical eye and the power to punish. All the sounds, sights, smells, and sentiments associated with religion begin to shape the consciousness of children before they know what's going on or can defend themselves. Audible calls to worship and prayer—church bells clanging, loudspeakers blaring calls to prayer from mosques—penetrate the minds of infants and children even while they sleep. Christian children become comforted with affirmations like "I've got a friend in Jesus" and tautologies such as "Jesus love me, this I know, for the Bible tells me so." Like a cuddly toy, the Baby Jesus becomes the archetype of innocence, goodness, and personal safety.

As children mature, dependence on parental authority changes to reliance on cultural institutions, especially religion, with its promises of physical protection and eternal salvation.[65] The psychological mechanisms and social pressures needed to promote the leap of faith have already been established and cultivated. For Christians, Santa Claus becomes God, the eagle-eyed observer and paramount judge of what people think and how they behave. Drawing on resources from their revered traditions, many Jews, Christians, and Muslims maintain this auto-infantilization for the rest of their lives. By passing along the parables, empty promises, and threats to their children, adults further harden their own commitment to the faith.

Organized religion gradually transformed the magical thinking and superstitions of the past into many of the dominant discourses, cultural identities, and moral codes of the present. What some call the god meme has no earthly parallel. Its power is rooted in our darkest fears—death, dying, illness, tragedy, the unknown—and in our brightest hopes—health, happiness, prosperity, and a vacation-like afterlife where, as comedian George Carlin quipped, we get to hang out forever with Mom, Dad, our grandparents, and our clean-living friends all in perfect health and in the prime of their lives.

Fear is an inherited emotion. We respond to threats, real or imagined, to make important decisions. A person walking down a dark street, for instance, might fear that someone is lurking in the shadows and preparing to attack. The person picks up the pace and arrives safely at the destination. The emotional reaction proved to be wrong—there was no stalker. But so what? Had there been an assailant, the person reasons, the faster pace probably would have foiled an attack. Exactly this kind of calculation was proposed by seventeenth-century French philosopher Blaise Pascal on the question of divine power. The formulation is known as "Pascal's wager."[66] Pascal wondered why anyone would believe in God. What are the rewards and costs of belief? Pascal answered the question this way: If you believe in God and there is no God, you lose nothing. But if you believe in God and are proven right, you gain everything. From this perspective, the dream of immortality costs little. Betting on God makes good sense.

But does evidence-free decision making about the existence of God really cost the individual little or nothing? Wouldn't the time invested in religious rituals, ceremonies, and everyday devotional practices be better spent doing things that contribute more directly to survival and reproduction? Richard Dawkins recalculates Pascal's wager in existential terms: "It could be said that you will lead a better, fuller life if you bet on his not existing than if you bet on his existing and therefore squander your precious time on worshiping him, sacrificing to him, fighting and dying for him."[67] The costs of believing in God don't just affect the individual. Doctrinal certainty and moral righteousness have been used to rationalize war, homicide and suicide, blatant discrimination against women and homosexuals, self-flagellation, and many other unethical acts.

The claim that (our) God embodies the absolute truth can be

found in all the monotheistic religions, but the infallibility conceit persists most strongly in Islam. With the exception of neuroscientist and author Sam Harris and a few others, Western commentators, including Richard Dawkins, shy away from specifically criticizing Islam, usually claiming they don't have sufficient personal familiarity with the religion to forge a comprehensive analysis. Ayaan Hirsi Ali filled that void when she became the most visible, thoughtful, and forceful public critic in the West. She narrates her break from Islam and escape from Somalia in her books *Infidel* and *Nomad* and in speeches across America and Europe. She complains that Muslims are taught to unquestioningly accept the claim of the Prophet's infallibility and to blindly follow an irrelevant and harmful ideology. Islamic believers lean more toward fundamentalism than do members of other religious groups, she says, because so many Muslims interpret the Koran as the literal word of God and have insufficient exposure to other religions and cultures.[68]

Some Muslim leaders and other intellectuals have taken a stand against terrorism, but they rarely fault the doctrines, requirements, or history of the faith. Like Hirsi Ali, most of those who speak out live in the West, where freedom of speech is a cherished value. In *The Trouble with Islam Today* and *Allah, Liberty and Love*, Canadian Muslim Irshad Manji imagines how the faith can be reformed from within. German political scientist Bassam Tibi's *The Challenge of Fundamentalism* critically dissects Islamic "ethno-fundamentalism." Bernard Lewis, a British American historian, wrote *What Went Wrong* to provide perspective on Islam's enduring problematic engagement with modernity. Sam Harris refuses cultural relativism in *The End of Faith*, and Christopher Hitchens is characteristically blunt about the perils of Islam in *God Is Not Great*. But volumes like Christopher Caldwell's *Reflections on the Revolution in Europe: Immigration, Islam, and the West*, Bruce Bawer's *While Europe Slept: How Radical Islam Is Destroying the West from Within*, Lee Harris's *The Suicide of Reason: Radical Islam's Threat to the West*, or Paul Berman's *The Flight of the Intellectuals* are often dismissed without serious consideration or convincing counterargument as paranoid, racist, or Islamophobic.

THE POWER OF RELIGIOUS COMMUNICATION

A belief system becomes influential when it is communicated persuasively to others. Discursive devices are required. In the case of religious communication, grand narratives passed from generation to generation turn the abstract God into particular beliefs and cultural practices that bridge the gap from fantasy to reality. Narratives are particularly effective teaching techniques for socializing children. The stories and characters give coherence, color, intrigue, and legitimacy to the confusing and often implausible historical events that make up the core claims and tenets of religious belief.

Most religious narratives begin with an inspirational protagonist. For committed Christians, a beautiful story is told of a humble carpenter, born of a virgin, sent by God to save the world, who performed miracles, died on a cross, and ascended triumphantly to heaven. Devout Muslims are charmed by the legend of an illiterate and ordinary man who was chosen by God to receive the final and true revelation and establish, by physical force if necessary, the rightful place on the earth for followers of the morally superior faith. Both prophets' influence derives from the way the highly stylized narratives bring the historical figures to life in an idealized way. The stories inspire hope through the appeal of creationist fantasies, miracles, selfless prophets, a loving personal God, and the promise of everlasting life. The poor and the illiterate are particularly vulnerable to the beauty of the words and the hope they portray.

The various creationist and "birth of humanity" tales expressed in religious mythology are deeply ethnocentric. The details that are described reflect the cultural worlds that existed where and when the stories were fabricated. The Garden of Eden, for example, is a clear projection of what an agricultural society might have dreamed up two thousand years ago. Middle Eastern cultures historically valued material objects that continue to represent great worth in ordinary life today—gold, precious stones, and herbal curatives among them.[69] Robert A. Hinde points out that "the Christian God is portrayed as sitting on a throne in a situation resembling a medieval court" in Bible stories and art.[70] Keeping with the theme of historical authenticity, many clerics and scholars today retell the religious legends as if they actually happened. Did Muhammad really go the mountain where he was embraced by God and given the final revelation? Did Jesus actually walk on the Sea of Galilee, give sight to the blind, and change water into wine? You might think so, listening to the way scores of contemporary commentators spin the myths.

The religious fables are poetic and magnetic. Believers sign on to their particular projects even though the stories—like the claim about the origin of the earth accepted by the Abrahamic faiths—are provably false. Creationist tales sound like primitive science fiction. A sacred Chinese legend contends that God Pan Gu floated for eighteen thousand years through a cosmic black egg until he smashed it open, created the universe, and separated heaven from the earth—a fantasy that fueled Chinese ethnic identity for generations. Members of North America's Havasupai tribe are still taught that Arizona's Grand Canyon is the birthplace of all humanity, despite genetic ancestry evidence that completely obliterates the validity of the claim.

Religious faith amounts to belief unmolested by facts. The viability of the belief system nonetheless depends on individuals continuing to accept the literal or metaphorical truth of the stories they're told and giving credence (and money) to the institutional storyteller. Religious texts become one-book solutions for nonreaders. The ideological power of religious authority descends from its pervasive presence in political governance. But religious hegemony requires more than the presence of ideology and an institutional infrastructure. The faithful have to continually confirm the legitimacy of their belief through their actions. Religious discourses offer consolation and a sense of the sublime to those who perform the obligatory practices, especially quoting passages from the sacred texts and displaying the correct manner of prayer and other forms of ritual participation. Creativity is put on full display because religion is such a powerful force. God and religion serve perpetually as enchanting resources for human expression.

Like narratives, rituals dramatize the content of religious belief systems. Rituals penetrate consciousness in much the same way children become socialized culturally through rhyming, repetitious songs, and games. Kneeling or bending in prayer, chanting, and the manipulation of prayer beads resemble adult compulsive behavior.[71] Public prayers, pilgrimages, holiday celebrations, group fasting, speaking in tongues, baptisms, spiritual singing, self-flagellation, and other ritualistic behavior reinforce the validity of the belief system and the value of community membership. Observed behavior communicates powerfully. Participating in rituals distinguishes believers from nonbelievers in a publicly visible way. The more active the participation and the more painful the suffering, the greater the reward. Many Muslims say their lives change profoundly after going on a pilgrimage to Mecca, for instance.[72] Catholic, Mormon, and evangelical rituals are likewise extreme in nature and strong in effect.

Although the true story of life on earth is absolutely magnificent, it's difficult for nature's honest narrators to compete with the supernaturalists' seductive folklore, reckless promises, and the claim that a kindly Creator lovingly made all that is beautiful in the world. Unvarnished nature—spectacular as it is—doesn't offer the same psychological assurances, superheroes, parables, holidays, or promises of immortality. But the truthfulness of the stories we tell ourselves and others does matter. A. C. Grayling distinguishes between "naturalists" and "supernaturalists."[73] When naturalists admire a snow-topped mountain, the delicate wings of a butterfly, or a baby's smile, they see and appreciate the natural world directly and honestly. Science can be summoned to explain in factual detail how these things evolved to become part of the natural realm, governed by nature's laws. But when supernaturalists admire nature's beauty, they programmatically assign responsibility to an empirically unknowable force to explain what they see. Nature's beauty just proves God exists. And when popular stories depict the worship of nature in and of itself, as, for example, the 2009 film *Avatar* did so spectacularly, many religious authorities condemn the narratives as blasphemy.[74]

COMMUNICATING RELIGIOUS ALTRUISM

Buttressing believers' claims of moral supremacy, religious narratives inspire faith-based altruism—the signaling of saintly qualities through the performance of benevolent acts. In the Christian story, Jesus of Nazareth was sent by his father to save everyone on the earth. The gift of the Son of God is uniquely valuable and was not given to anyone in particular. The Crucifixion was an unrequested and unwanted sacrifice made by Jesus Christ on behalf of all humankind. The gift has universal recipients and can't be turned down. The overwhelming debt owed by humankind is unavoidable. The offering represents the design of a brilliant donor: create the ultimate obligation—a debt that was never asked for and can never be repaid. Humankind owes everything to the Creator, his Son, and to all those who claim to speak of his plan for humankind, to talk on his behalf about moral issues, and to collect his dues. Claims made in the narrative cannot be verified, and God cannot be consulted to confirm his intentions. The loving Christian God exhibits imperialistic intent: everyone should be shamed into becoming a Christian.[75]

Showing absolute faith in holy ghosts indicates personal greatness. But doesn't any demonstration of faith extract a high cost of time and energy on the part of the altruist—an evolutionary disadvantage? Yes, but by paying the price, the faithful reap the practical benefits that accrue from group membership while enhancing their individual status in the community. If an individual acts selflessly and virtuously by generously donating money to charity, for instance, the contributions elevate the giver's reputation in the eyes of the community. Whatever evolutionary costs incurred by the forfeiture of money, time, or energy are compensated for by the demonstration of what the person can afford to give up for God. Anonymity only enhances the greatness of the donor.[76] The giver, too, experiences warm self-gratification.

What attracts people to this kind of giving is a promise of the ultimate reward for moral behavior: immortality and the transport of an unproven entity—the soul—to a realm that has no beginning or end. The entitlements and penalties associated with such moral posturing are enormous. The more you give, the more you get back. Sacrifice your life for the good of the group, and the rewards can be boundless—ninety-two dark-eyed virgins were promised to each of God's warriors who knocked down the Twin Towers.[77] Eternal punishment awaits the selfish and the sinful.

This is phony altruism run amok. Religiosity is not sheer illusion, as Freud suspected.[78] Believers walk with utter pride among the infidels. God spoke with Jesus. Moses spoke with God. So did Muhammad. Even George W. Bush thought God told him what to do. Half the population of any drug or alcohol rehabilitation unit says the same thing. No wonder religions become breeding grounds of ignorance, intolerance, and war. Ecumenicalism can become a venomous ploy, an insincere piety, a discriminatory tool used to identify the enemy. In God's name, nature's altruism can be turned into an instrument of hatefulness and harm.

Despite the trumpeted emphasis on moral behavior and self-denial, religiosity depends on a level of selfishness and extravagance that most religious leaders and followers refuse to recognize. Religion's exclusivity assures the faithful they are worthy. The promise of everlasting life makes them immortal. "Most who claim to serve God actually want God to serve them," says researcher Gregory S. Paul, who argues that self-gratification and materialism represent the true universal condition and that religion just conforms to this underlying truth.[79] But in the minds of many, religious belief and material consumption complement each other unproblematically. Faith itself functions as social capital.

A display of religious fervor becomes a strategic choice—a way to identify oneself as a morally upright person. The high-minded status must be communicated to others. That's why fundamentalists wave their banner of faith the same way a brilliantly colored bird shows its feathers, the pyramids advertise the might of the pharaohs, or a pop star flashes the paparazzi. Akin to vanity, immodesty, and ornamentation, religiosity shines like a laser beam on the believer while condemning doubters to the shadows. An uncharitable truth becomes conveniently obscured by the misleading claim of selfless altruism. The perfect strategy is the one that doesn't speak its name.

This is not to say that altruistic acts performed in the name of God or under the auspices of any other authority benefit only the altruist or are wholly dishonorable. Religious workers and other volunteers tirelessly serve needy populations in Africa, Asia, Latin America, and the Middle East sometimes at great personal cost—even death. In societies that suffer from autocratic rule or where people struggle just to survive, religious organizations often provide needed social services that undemocratic, corrupt, or impoverished states fail to offer. The Muslim Brotherhood's role during Hosni Mubarak's reign over Egypt from 1981 to 2011 is a recent example. Hezbollah and Hamas deliver the same services in Lebanon, Syria, and Palestine. These political charities function as group-level adaptations to stressful conditions and demonstrate a basic evolutionary principle: we fare better as individuals and communities when we behave kindly toward others.

Religion performs best as a social and cultural force in societies in which nearly everyone is devout, in which everyday life is organized around pietistic traditions, and in which the rates of education and literacy are low.[80] Most of these societies are poor. Parsing out the causes and effects of social misery is delicate work, yet the impact of religious belief on cultural development must be honestly examined. Although he hesitated to criticize religious beliefs directly, Charles Darwin observed that "within the less civilized nations reason often errs, and many bad customs and base superstitions come within the same scope, and are then esteemed as high virtues and their breach as heavy crimes."[81]

Scholars typically blame the conditions of underperforming nations today on political corruption, economic underdevelopment, illiteracy, urbanization, and population demographics, while Judaism, Christianity, and Islam are celebrated as "the world's great religions." Journalists and media commentators avoid examining the actual origins and consequences of the customs and superstitions Darwin described. Instead of holding the proponents of religious intolerance responsible for death and destruction in troubled parts of the world, we get euphemisms like "interethnic differences," "civil war," and "sectarian violence." Politicians dance clumsily around the issue. Social media sites prohibit or censor religious criticism. No one likes to point the finger at someone else's spiritual beliefs and practices.

Blasphemy threatens personal identity and social stability. A political factor is at work, too. Blasphemy is counter-hegemonic; it endangers elite individuals and institutions that wield cultural power. Even in comparatively developed countries, the "sacred" receives special consideration. When religionists speak, the content of their message is "belief." Any attack on religious belief is considered to be personal, mean-spirited, and possibly illegal. Even mild religious satire can be interpreted as "hate speech." In contrast, when secularists express a nonreligious point of view on an issue, the message is regarded as mere opinion, not sacred belief. No special sensitivity needs to be given to the speaker nor protection to the message.[82]

The religious right and the secular left tend to agree on this kind of censorship. This thinking is misguided. In a free society, no person or institution has the right not to be offended. For a society to develop healthily, the sacred, like everything else, must be open to sharp and thoroughgoing criticism. To argue otherwise is to concede that some cultural power brokers need not be held accountable for their doctrines or their behavior.

AN OPENING

In *Why I Am Not a Christian*, a well-known collection of essays published in 1957, Bertrand Russell said he believed science had already produced enough knowledge to secure universal happiness and well-being. "It is possible that the world is on the threshold of a golden age," Russell wrote, "but, if so, it will be necessary to slay the dragon that guards the door, and this dragon is religion."[83]

The dragon will not die anytime soon. It works too well for too many people. And Russell was overly optimistic about science, which often creates problems faster than it can solve them. But his contempt for religion is well-founded. How can universal happiness and well-being ever come about when differences based in religious traditions—each with its exclusive access to God and claim of moral supremacy—keep people apart? We've seen some of the worst material effects this kind of thinking can deliver. But in the broader sense, fundamentalist discourse is more of a problem than terrorism is. Fundamentalist religious principles infect far more people than make up the radical fringe. That's why Richard Dawkins says he does "everything in [his] power to warn people against faith itself, not just against so-called 'extremist' faith."[84] He writes, "Moderate religion makes the world safe for extremist religion by teaching that religious faith is a virtue, and by the immunity to criticism that religions enjoy."[85]

While the dragon of religion may never be slain, it is steadily weakening in the Communication Age. Popular commitment to all Christian denominations has declined across Europe for years. The Nordic and Baltic countries have become particularly nonreligious. England, France, the Czech Republic, and the Netherlands strongly fit the trend.[86] The majority of all Britons no longer belong to any religious group.[87] The lone Muslim-dominant state in Europe, Turkey, is by far the most religious on the continent, although, by Islamic standards, the degree of religious commitment is low there, too. Only the Christian orthodox churches of Romania, Macedonia, and Georgia remain relatively influential.[88]

Widespread disillusionment with the Catholic Church's teachings and sex scandals has caused many parishioners in the United States to leave the fold or to change denominations. More than 70 percent of the American public now believes religion is losing influence.[89] The religious exceptionalism that has been so much a part of the nation's history is gradually giving way to free thinking about spiritual matters.[90] Far fewer Americans today associate themselves with any single faith, and many others look for spiritual guidance from nontraditional sources. Tepid believers, opponents of organized religion, agnostics, and atheists outnumber Muslims, Jews, Mormons, Jehovah's Witnesses, Seventh-Day Adventists, Buddhists, and Hindus combined in the United States. The number of Americans who believe in "secular evolution" has nearly doubled in recent decades.[91]

These transformations fit within a broad pattern of cultural change in America. Many people in the United States claim to be Christian, when what they are really saying is that they are Americans. For these individuals, "Christian" is more a cultural than religious designation. They unthinkingly identify themselves as Christian because that's how they were raised. Going to church was just a family routine. They may abstractly believe in God but are not necessarily religious. The same can be said for many Jews and Muslims in America.

Myriad alternatives to traditional religious faith are cropping up. More and more people piece together do-it-yourself cultural experiences and spiritual identities.[92] They integrate reincarnation, yoga, and other Eastern teachings and mystical musings like astrology and numerology into their worldviews. Some include natural forms of spiritualism—reverence for mountains, oceans, trees, and crystals, for instance—into the mix. Even the majority of Americans who continue to identify with a religious denomination no longer believe their faith provides the only path to happiness or eternal life. Mainstream (non-Evangelical) Protestants and Jews are most open-minded in this regard, but slightly more than half of the American Muslim population also agrees.

Support for fundamentalist ideology and terrorism has been declining among Muslims in most parts of the world in recent years.[93] Support for suicide bombing is falling off. Islamic terrorist group Al-Qaeda is viewed unfavorably by majorities in Jordan, Lebanon, Indonesia, Egypt, Pakistan, and Turkey. As a rule, modernity diminishes religion's influence. Except for Egypt, most Muslims in these countries identify more with modernizers than with religious conservatives. Majorities in all but Pakistan think democracy is preferable to any other kind of government. Muslim publics generally support educating girls and boys equally.[94]

Women are leading the way for reform in the Islamic world. Communications media provoke and sustain the discussions. The first experience that encouraged Ayaan Hirsi Ali's escape from Somalia, for example, was reading the "tales of freedom, adventure, of equality between boys and girls, trust, and friendship" she found in Nancy Drew mysteries and Enid Blyton's children's novels.[95] The Arab world's most popular television program, *Noor*, from Turkey, is a serial about an uncovered Muslim woman with an open mind and a blue-eyed, blond boyfriend.[96] Speaking to the needs and passions of the universal woman, Oprah Winfrey's television program has been on the air in the Middle East since 2004. Jordan's Queen Rania continues to promote Islamic modernity. Rape victim Mukhtar Mai's much-publicized struggle for women's rights in Pakistan symbolizes cultural change in south Asia. The image of Neda Soltan lying in a pool of blood, shot to death in the wake of Iran's presidential elections in 2009 on the streets of Tehran, still sparks political unrest in the Islamic Republic where she's known as *Neda Iran*—the "voice of Iran." Eman al-Obeidy received worldwide media coverage when she stormed into a hotel where international journalists were staying in Tripoli to publicly reveal the physical abuse she received for supporting the Libyan resistance. Police in Saudi Arabia arrested Manal al-Sharif for posting a Facebook® and YouTube® image of herself driving a car as part of a spirited protest for women's rights.

Historian of technology George Basalla notes that "the Arabic word *bid'a* has a double meaning: 'novelty' and 'heresy.' The worst kind of *bid'a* is imitation of the ways of infidels. The prophet Muhammad warned, 'Whoever imitates a people becomes one of them.'"[97] In a profound sense, Muhammad was right. Cultural behavior necessarily changes when people imitate the ideas of their neighbors and visitors.[98] To the righteous indignation of cultural traditionalists everywhere, good ideas—like adaptive biological mutations—usually win out over time. For example, most modern societies have begun to recognize the legal rights of gay people, to allow abortions, to encourage birth control, to guarantee free speech, and to ban religious infringements on public education.

Rigidity is a losing evolutionary posture. Even religions are forced to adapt to new realities when common goals that can override the benefits deriving from sectarian ideology are established in a society.[99] But common goals can be established only when information circulates widely. New ideas—like biological mutations—have a chance at life only when the environment provides sufficient breathing room. Communication is the social process through which cultural change takes place and is further expedited by the latest developments in technology. Communications media are inherently adaptive. Early language, chanting, cave paintings, and other simple media helped primitive peoples forge tribal bonds that worked to their evolutionary advantage. Literacy and print technology later spurred the development of religious and secular alternatives, another improvement. The first true mass medium—the printing press—empowered Martin Luther's critique of Catholic hegemony in the Middle Ages, triggering the Protestant Reformation and setting Western culture off on a new path. The free dissemination of ideas lay at the heart of emerging Enlightenment values and served as a precursor to political and cultural modernization. Freedom of speech later became a constitutional guarantee in America, a human right that has been adopted by many other countries and by the United Nations.

ACCOUNTING FOR NATURE

Nature's magnificence is so overwhelming that humans have always tended to attribute its creation to some mysterious celestial power. Lacking any deep understanding of the natural world, early humans invented gods of the sun, earth, sky, oceans, rivers, rain, thunderstorms, earthquakes, volcanoes, love, war, and various diseases.[100] The power of pagan gods and goddesses sprang from the forests, rivers, animals, and other natural forces where the myths originate. The Greeks considered Zeus, the god of the sky, to be the king of gods and attributed the outbreak of plagues to divine disapproval. Nature's splendor is proudly celebrated as proof of God's existence by Jews, Christians, Muslims, and Buddhists. God told Noah that he would create rainbows to represent his covenant with man, birds, livestock, and the various wild animals that were crammed onto the ark.[101] According to sura 6, verse 59 of the Koran, all of nature is known to God right down to the tiniest earthly detail: "He knows what is in the land and sea; not a leaf falls, but He knows it." Sitting beneath an ancient fig tree, the Buddha believed he began to understand the natural world when his hand caressed the earth. And when he died, trees burst into bloom.

Of the religions, the Buddha's "supreme and final wisdom" comes closest to the truth of evolution.[102] All of nature is connected. Life is about impermanence and flow. Whatever is born is subject to decay. No afterlife should be expected. The spiritual only mimics what nature does. But religious pontification of any kind can only be hopelessly reductionist and misleading. The evidence isn't there, and the scale is completely wrong. Recent scientific advances show that multiple universes composed of billions of solar systems continue to be formed. The systems of the multiverse derive from physical laws and require no divine intervention. No less an authority than Stephen Hawking has finally made clear that the romantic idea that the earth was created by a supernatural being or god for the benefit of humans has no credibility whatsoever.[103]

CHAPTER 7

COMMUNICATING CHANGE

More than one-third of the entire population of England, some six million people, paid one shilling each to attend "The Great Exhibition of the Works of Industry of All Nations" in London's Hyde Park during the summer of 1851. Charles Darwin was among them. Like all visitors, Darwin was surely impressed by the sparkling structure that housed the exhibition—the majestic Crystal Palace. Once inside, he would have strolled up and down the aisles, admiring more than a thousand booths that celebrated the growing achievements of nineteenth-century science and technology in chemistry, metallurgy, manufacturing, horticulture, commerce, glasswork, machinery, architecture, agriculture, and many other industries. The young field of technical communications was represented with displays of advances in printing, telegraphy, and photography. Even a prototype of the fax machine was demonstrated. The stereoscope—which for the first time permitted realistic three-dimensional viewing of an object—was a particularly sensational exhibit with research applications that must have intrigued Darwin.

The Great Exhibition served as the cultural coronation of the Industrial Revolution. The fair represented a tremendous source of wealth and prosperity and a rosy future for Britons. It marked the beginning of what one author called "planetary capitalism."[1] Great Britain was fast becoming the economic leader of the newly industrializing nations, a pioneer in the development of science and technology, and the world's most successful imperial power. All this was evident by the range of impressive industrial output that originated in British colonies—especially in India, Australia, and New Zealand. Communications technology that could move information quickly around the globe was becoming crucial to England's imperialist exploits and to its domestic economic growth.

Darwin attended the Great Exhibition eight years before he published *On the Origin of Species*. Much of what he observed around him during the book's gestation period, including his experience attending the international fair, confirmed the line of scientific reasoning he had been developing since the *Beagle* voyage. Nature and industry were proving not to be at odds with each other; both discredited any static view of life. A spirit of inventiveness appeared everywhere. Patents for inventions grew at a phenomenal rate.[2] Technological breakthroughs were taking place through the ingenuity and hard work of inventors and craftspeople who had no conventional schooling in hierarchical Great Britain. Practical knowledge mattered like never before.

Great Britain had become the driving force of nineteenth-century modernity and globalization. It would be a mistake, however, to interpret the significance of the Industrial Revolution mainly in terms of economic growth and the practical application of technology to commercial production. Innovativeness and openness were developing into cultural values to the point that the very basis of traditional society was being dismantled. The quantity and quality of scientific evidence and technological power on display at the Great Exhibit severely contradicted the ideals of Victorian culture. The material world was exploding productively at the hand of humans, not God, and the locus of political power was shifting from kinship networks to the authority of the state. Core Enlightenment values—secularism, the idea of progress, and domination of nature—were taking hold.[3] Literacy and education were on the rise. Modern journalism was born. Enormous libraries and archives were being established.

No beginning or ending date for British industrialization can be declared, and the technological advances that were made proceeded incrementally. No radical break from evolutionary patterns had occurred; everything was just speeding up in response to new human-made environmental conditions. Steady industrial growth and a global mindset emerged as hallmarks of the period.

THE NATURE OF INDUSTRY AND TECHNOLOGY

A positive correspondence between biological evolution and technological development was becoming clear. Evolutionary principles that underpin biological heredity and hybridity could be found in the openness, complexity, and malleability of culture. Just as biological organisms diversify from a common origin shaped by natural selection, technological development spins a web of complexity from more simple ways of life shaped by culture. Favorable natural environments increase the production of biological mutations, heightening the chances that superior variations will emerge. Cultural environments that encourage innovation and the free flow of information prompt technical development that can bring about transformational results. The repercussions are striking. Early technological development, like the printing press or the compass of the sixteenth century, leads to the creation of additional innovations and predicts the level of economic development for as long as five centuries later.[4]

All life systems generate their own momentum,[5] but conditions that allow for experimentation must be present for the pace of change to increase. From the very beginning of our planet's history, the "watery origins of life" allowed for the movement and collision of elements that could spark life and form biological systems.[6] That is precisely what was happening culturally in the freewheeling mid-nineteenth century. Industrial technology was becoming the new nature; machines, the symbol of procreation. Driven by the force of the human imagination, the fruits of the Enlightenment were maturing in material and cultural form.

Galloping industrial production established new economic priorities and practices in Great Britain and in the rest of the modernizing world. European industrialists were giddy over their prospects. To some critics, however, the consequences seemed ominous. Upstart industries live on the brink of financial disaster. Greater and greater amounts of capital would have to be invested in machinery, factories, and transportation. To pay for all this, wages would have to be depressed, putting unskilled and unorganized workers at high risk. Low pay also meant that the pool of consumers would likely stagnate or shrink. Production could outpace consumption. If that happened, the overproduction of unsold goods might lead to a recession or depression that could destroy the economic strength of the industrial barons and their backers—including the investment-savvy Charles Darwin.

The solution to the potential crisis was to grow the market. Colonial expansion was encouraged by industry and government. People responded in droves. Between 1846 and 1890, the number of people leaving Europe rose to 377,000 yearly, increasing to 911,000 emigrants annually from 1891 to 1910.[7] Technological advances that led to fast-rising income levels and robust patterns of emigration defused Thomas Malthus's dire warning that without abstinence and family planning, population growth would bring social disaster to Great Britain.[8] During the Industrial Revolution, some 35 percent of Great Britain's manufactured goods were sent overseas. British exports climbed to 46 percent of the world's textile production.[9] Offshore industrial production was initiated. Laws that restricted the export of machinery to the colonies were suspended.[10]

The way forward had been paved by the Europeans' "discovery" of other continents, their colonization of new lands, their exploitation of foreign resources and laborers, and the massive export of human capital to every corner of the earth. Foreign markets sprouted on the heels of the migrations. The dream of creating a world market for industrial products from Great Britain and continental Europe was becoming a reality. International trade boosted consumer activity.

English, French, and Spanish were becoming world languages. A global economic system favorable to British and other European manufacturers had been set in motion. The West was ascending as the economic and cultural power center of an increasingly interconnected world.

Nineteenth-century capitalists formed the first ruling class to identify with the irreverent idea of a dynamic, forward-looking, secular society driven by relentless technological change.[11] Exactly what technological development and industrialization would ultimately mean in social and cultural terms, however, was not clear. Implicit evolutionary explanations were offered. Karl Marx, for example, granted that although the industrializing world would assuredly bring about negative consequences for the working class, the vitality of industrial production emerges innately from the need to survive. The parallel between biological and industrial processes was clear to Marx.

He saw a close connection between the "living" and "made" worlds, as he called them—between the physical organs biological beings need to exist and the tools and other artifacts humans create to both improve their lives and to ensure continuation of their lives. Dependence of the human body on nature explains why humans are driven to explore cultural possibilities by creating diverse and sophisticated technology, Marx thought.[12] The body uses nature for biological purposes just as labor turns natural resources into material objects. Natural and sexual selection produce tangible results the same way human labor does.

Karl Marx and Friedrich Engels also anticipated the importance of expanded social communication. Consciousness and culture were being transformed in the industrializing world. "National one-sidedness and narrow-mindedness become more and more impossible," they wrote in *The Communist Manifesto* (1848), "and from the numerous national and local literatures, there arises a world literature."[13] What they meant by "world literature" developed into what we now call global media culture. Originally appropriated by European economic power brokers to exploit the growing global economic market, early industrialization brought with it "immensely facilitated means of communication," according to Marx and Engels.[14] Every technological and cultural nuance added a new link to the ever-expanding chain of communication. The Industrial Age was becoming the precursor to the technological prowess and cultural vitality that defines today's Communication Age.

In many respects, the assumptions that underlie Marx's economic and social theory parallel the basic principles of evolutionary theory that Darwin was conjecturing at roughly the same time. But so, too, did the opposing theories of social philosopher Adam Smith, with which Darwin was also well acquainted.[15] Smith's foundational economic concepts—specialization; free trade; entrepreneurship; and the power of a vibrant, competitive market—have their analogues in nature, too. The division of labor and market tendencies that curb unrestrained greed—corrective mechanisms championed by Smith—could be seen to operate even among lower life-forms. Despite the profound differences between the two men, Marx and Smith both agreed with Darwin on the most critical point: fundamentally, life unfolds as fierce competition among self-interested beings.

Rampant industrial growth helped Darwin understand the natural world. Biographer Janet Browne points out that when Darwin wrote *On the Origin of Species*, he "drew on industrialized England for a metaphor. Natural selection probably favored those animals and plants that diversify just as if nature were a factory bench in which production was more efficient if workers performed different tasks."[16] Biological organisms that diversify—mutate into variations that adapt well to their environments—survive. Institutions that diversify—employ people with varying talents and create products and organizational structures that reflect changing market conditions—prosper.[17] Diversity of biological species results from random variation in nature; the process is undetermined and self-sustaining. Diversity of ideas and commodities results from entrepreneurial efforts in the technological, industrial, and cultural arenas; the process is goal oriented and driven by human agents.

If the presence of diversity demonstrates successful evolutionary outcomes in nature, what are the necessary antecedents of *biological diversity*? An abundance of random mutations, adaptations worked on by natural selection, and an array of consequent variations. And if *artifactual diversity* produced by technological advances serves human societies well, what conditions must be present to fulfill that potential? Curiosity, a spirit of innovation and entrepreneurship, and the freedom to make choices. Technological innovation creates superior material forms while destroying outmoded industries and products just as new biological species emerge in response to changing environments while stagnating species recede or disappear.[18] Like the successful adaptive outcomes of biological evolution, technological development affirms the human potential. As George Basalla describes it, the history of technology "is a testimony to the fertility of the contriving mind and to the multitudinous ways that the peoples of the Earth have chosen to live. Seen in this light, artifactual diversity is one of the highest expressions of human existence."[19]

Nature's tremendous diversity can be explained only by the gradual dissemination of animals and plants across the earth's entire broad surface. Organisms arriving in new geographical territories spur physical modifications over time, eventually leading to the creation of new varieties and species. As Charles Darwin pointed out, even the unconscious "occasional transport" of seedlings caught in the feet, legs, and feces of migrating birds would give rise to the multiplication and modification of countless species of plants around the world.[20] Change results from adaptation. Adaptation requires mobility. Biological beings, including early humans, were driven to expand their geographical horizons to survive. The spread of technology and industry follows the same basic pattern. From the biological beginning to the technological present, evolution is a thoroughly globalized phenomenon.

Historical parallels between the way industrialized capitalist societies developed to produce commodities and how they grew to become global producers and distributors of messages are striking.

In preindustrial societies, people had to be extremely self-reliant to raise food, to construct and maintain living spaces, to make and repair clothes, and to perform most all other details of everyday life. Because some individuals proved to be more skilled than others at carrying out certain tasks, preindustrial specialization—the age of the artisan—emerged. The first forms of organized industrial activity, early manufacturing, soon followed. Men and women with various talents applied their trade for bosses who managed production, cultivated markets, and sold the first mass-produced goods. Full-blown industrialization grew from the early manufacturing stage. Assembly lines accelerated the speed and efficiency of production, upstart corporate structures mushroomed, and marketing and advertising were developed as integral related industries. The production, marketing, and trade of industrial and consumer goods gradually became less tied to physical location and local ownership as economic globalization grew to enormous proportions by the start of the twenty-first century.

A similar trajectory can be traced in the production and spread of information and entertainment. The forms of social communication characteristic of emerging modern societies were inherited from our ancestors—routine, unmediated interactions that took place in families, neighborhoods, and communities. But much like the artisan's role in the production of commodities in early modern societies, individuals with the talent, desire, and opportunity could become de facto information and entertainment specialists—storytellers, teachers, preachers, musicians, artists, writers, and orators. The further development of communications technology in capitalist societies led to the mass production of messages in much the same way that machines and the organization of labor made the expanded manufacture and distribution of commodities possible. New varieties of public communication were instigated and financed by business agents and independent producers (to negotiate contracts with publishers or to arrange musical or theatrical tours, for example). As print, film, and electronic media technologies emerged in the late nineteenth and early twentieth centuries, public information and entertainment became increasingly commoditized and commercialized. The production and sales of news and popular entertainment became lucrative industries, corresponding historically to the industrialization of consumer goods. The constituent elements and forms of power that contributed to economic globalization were soon matched by the ubiquity and force of mass media, the Internet, personal communications technology, and the transnational culture industries.

Improved communication efficacy has always strengthened our species' prospects for survival. That's why there's been such a sharp focus throughout history on creating technologies that make it easier to connect with others. The evolution of contemporary communications technology has driven global modernity for more than five centuries because useful innovations inevitably spread across social groups and cultures. Wide familiarity with dominant languages and technologies facilitate the diffusion of innovations globally. End users determine the innovation's success, and the personal benefits of most new communications technologies for the individual are tremendous. The new global individual thrives as a superior communicator and culture producer replicating to infinity.

The macro consequences of technology's global reach are equally noticeable and important. Political systems tend to become more democratic as communications technology evolves. Education and literacy increase. Societies become more tolerant and progressive overall. The middle class grows larger. Vital health information circulates more widely. Physical mobility—forms of transportation—expands and improves.

The positive trends confront formidable countervailing influences as well. Chief among them are the following:

- the sheer scope of the global information landscape that creates a sense of shock, helplessness, uncertainty, suspicion, and distrust for some people;
- individuals may make less effective decisions because the constant arrival of information interrupts sustained analytical thought;

- censorship by state and religious authorities in various nations limits the free flow and reception of ideas and information;
- corporations disproportionately influence the production of media content;
- the contamination of global communication by the spread of political and religious propaganda and disinformation, including terrorism;
- the sensationalist tendencies of media, the culture industries, and contributors to web content;
- the digital divide between and within nations and cultural groups, and the underlying economic and educational gaps that influence differential access to and use of communications technology;
- excessive self-interest emerging from the availability of more diverse media content and cultural offerings, and the personalizing features of information and communications technology; and
- strong tendencies by people to seek, avoid, perceive, and retain ideas, information, and experiences that accord with their cultural socialization sometimes leading them to retreat into a tribal mindset.[21]

SHAPING EVOLUTION

Nature is a random tinkerer. Selection acts on random mutations to produce outcomes that reflect conditions present in diverse environments. The mutations that survive and flourish are the ones that adapt well to these situations, a process that unfolds without fixed plans or intentions and that does not always lead to superior solutions. But while adaptations always increase the fitness of the individual, they don't necessarily improve the overall qualities of the species.[22]

Things change, however, when humans intervene. Humans can influence the course of nature's production by altering the biological or cultural environment in which the production takes place. Random and nonrandom tinkering both represent the essential process that underlies all of nature's manifold production—organic evolution. To gently convince skeptical readers of this discomfiting fact, Darwin began *On the Origin of Species* by describing a kind of biological production with which the layperson was already familiar—the way domestic breeders of horses and pigeons vary their species' offspring by controlling their mating patterns. He detailed how purposeful domestic breeding—artificial selection—proceeds in a manner very similar to the blind and dumb tinkering that goes on in the natural world, except that domestic breeding is humanly guided. The technological innovations and industrial production that blossomed so spectacularly during Darwin's day extended the principles of biological evolution to the manufacture of material artifacts.

People reflect on and meddle with their own biological potential, too, even with their genetic inheritance. Nazi Germany's experiment in social engineering, the breeding of slave populations, and the idea of human cloning are the archetypal examples. Many ordinary individuals attempt to engineer their own social relationships by commending or arranging marriages; ostracizing undesirable mates; aborting unwanted births (sometimes according to gender or race); bullying ethnic, sexual, and religious minorities while embracing other groups; shunning disabled persons; and so on. These behaviors are not determined by evolution, nor do they in any way represent evolutionary theory. They are human decisions rendered from a range of alternatives.

Inclusive, positive cultural development can be humanly guided, too. For Darwin, the idea that nature has certain determining properties did not mean civility could not be learned. Darwin allowed that people can take "some pride" in evolution because of "man's powers of sympathy, benevolence, and intellect."[23] Evolution does not equate to vulgar self-interest. Even the basic instinct to reproduce without restraint has been controlled. Brute natural and sexual selection do not determine our species' future; the self-aware, other-directed human being can do better than that. Our fate depends more on how we adapt to the environment—something over which we have significant control. Apart from the obvious cultural challenges, environmental issues, including the ravages of climate change, population control, water management, food safety, curbing disease, and protecting biodiversity, can be confronted productively because people have the capacity to choose rationally among competing courses of action. It is to our advantage to do so, and we've generally followed that course over the millennia with successful results.

Although doing so has always been challenging, human aggressive tendencies should be humanly manageable. The trend is going in the right direction; the idea of a peaceful past is a popular but misleading myth.[24] Many primitive societies were constantly killing each other, and cannibalism was widespread. Religious wars and colonization depended on superior firepower. Technological development and industrial growth greatly increased the sophistication and brute power of destructive weaponry. But despite the accumulation of firepower, human societies have gradually become much less violent than their premodern ancestors. During just the last two centuries, we've seen far fewer wars, a dramatic reduction in casualties caused by the wars that have been fought, less genocide (even counting Rwanda, Bosnia, Darfur, Libya, Syria, and recent atrocities in other regions), fewer arms deals, fewer refugees, the longest periods of peace between major powers, and fewer international crises overall.[25] Urban dwellers in modern countries today have never been safer. As economist Paul Seabright points out, the psychology of everyday life in the modern world makes it uniquely possible to "step out of the front door of a suburban house and disappear into a city of ten million strangers."[26]

Moral development and political democracy have also been on the move.[27] Formal recognition of basic human rights and property rights, greater gender equality, the eradication of diseases, vast improvements in literacy and education, increased tolerance of differing sexual orientations, reduced infant mortality, a dramatic increase in life expectancy, and many other accomplishments are evident. Scientific rationalism, the vitality of capital markets, and the information-sharing power of modern transportation and communication systems create conditions where civilizations can prosper.[28] Increased integration into the world economy, expansion of the private sector, the rise of the middle class, and greater entrepreneurial freedom promise to eventually transform even the very last of modernity's holdouts.[29]

Still, the generally uplifting record of human evolution does not mean peace and progress are inevitable. Right along with reducing violence and promoting human rights, we have also created the means with which to blast each other off the planet and destroy the earth's ecosystem. People empathize with and behave kindly toward each other, but they also retain the capacity for unspeakable violence and cruelty. This discordant reality is primordial. Aggression and kindness have deep evolutionary roots that show up in the behavior of other primates as well, especially in our closest genetic relatives, bonobos and chimpanzees.[30] These primates fight ferociously but also groom each other tenderly. Psychologist Steven Pinker summarizes the nature of the behavioral opposition: "Left to their own devices, humans will not fall into a state of peaceful cooperation, but nor do they have a thirst for blood that must be regularly slaked. Human nature accommodates motives that impel us to violence, like predation, dominance, and vengeance, but also motives that—under the right circumstances—impel us toward peace, like compassion, fairness, self-control, and reason."[31]

Natural and sexual selection reward both tendencies. The aggressive side of human behavior, including fear of others' aggressive actions, derives from instincts that were instilled in the primate brain long ago. Individual and organized violence can still provide reproductive advantage for their perpetrators. Domestic abusers, tribes, gangs, terrorists, and warring nation-states all enforce their identities and agendas with violence. Cruelty to animals for pure entertainment purposes takes place, usually illegally. Fascination with media violence surfaces in genres ranging from the brutal combat of Xtreme

Fighting on television and the mayhem of interactive video games to the anonymous bombing and shelling presented in war documentaries. When the ratings for the American wildlife channel *Animal Planet* started to slip, programmers at the cable outlet decided to emphasize "predation programming"—animal death action shows.

Kindness evolved within the broad framework of morality, but an inherently distrusting tribal mindset shaped the way kindness would be expressed. "Social virtues," Darwin wrote, "are practiced almost exclusively to men of the same tribe."[32] In-group and out-group distinctions that emerged in our evolutionary past instinctively predispose positive activity inward. Acts of altruism, generosity, empathy, and pity still tend to most benefit those persons who have the potential to advance the genetic interests of the altruist—close kin and potential replicators.[33] Tribal instincts that limit social cooperation remained in place even as societies became larger and more complex.[34]

Those constraints began to loosen, however, when our ancestral groups came into more frequent contact with each other and started to exchange goods. Our species' peculiar predilection to innovate ignited the positive development. The invention of stone hand tools—a crucial evolutionary development marking the birth of technology—encouraged cooperative behavior between tribes. Tools found fifty kilometers from the place of their origin suggest that intertribal trading networks may have emerged some two hundred thousand years ago, long before many of our ancestors headed north.[35] The interactors, strangers to one another, began to treat each other as distant kin or community members.[36]

Technology, trade, and social cooperation evolved together and were empowered by improving communication ability. Simple barter and exchange became the key mechanisms for the development of human civilizations and represent the first instances of modern intercultural communication. In the process, a preference for negotiation over annihilation may have seeped into the human genome, progressively dulling the sharp edge of cultural confrontation.[37] More elaborate forms of social reciprocity evolved through a long history of economic transactions that were facilitated by the cooperative spirit of symbolic exchange.

PRO-SOCIAL COMMUNICATION

Our natural tendency to take care of people we know also allows us to empathize with individuals we don't know or those who would not be able to return any favor. Like the instinct for self-protection that led to tribalism, the instinct to empathize runs deep. People feel sad when they see other humans, animals, even insects suffer. Suffering encourages generosity among unknowns, as is demonstrated by people donating to charitable telethons (a universal television genre), or responding to pleas of support for the victims of tragedies, especially if the victims are poor or disabled. The ability to transfer empathetic feelings from known to unknown individuals inside and outside our species is a by-product of behavior that was cultivated long ago.

Communications media can intensify and extend the pro-social effect, even when the narratives are fictional. Novel reading in the eighteenth century, for instance, helped people develop emotional connections with each other, confirm the shared nature of their inner feelings, and prepare the ground for raising global consciousness about human rights.[38] We cry over fictional characters at the movies. People recoil when they see videos or photographs of people suffering, fictional or real. Cable television's evocative coverage of the famine in Ethiopia in the 1980s "created a new kind of electronic internationalism linking the consciences of the rich and the needs of the poor."[39] Today's social media allow anyone with Internet access to circulate ideas and images that can trigger empathetic responses by unknown others.

Timeless biological principles and modern communication practices combine to create a psychological evolutionary substratum that supports and encourages constructive human behavior. Violence is reduced, discourses for cooperation are created, a platform for moral consensus is formed, and empathy spreads beyond cultural borders. Darwin recognized that many positive social dimensions of our evolutionary legacy are also shared by other species: "Social instincts lead an animal to take pleasure in the society of its fellows, feel a certain amount of sympathy with them and to perform various services for them."[40] Darwin's interpretation of human history was clearly hopeful: "As man advances in civilization, and small tribes are united into larger communities, the simplest reason would tell each individual that he ought to extend his social instincts and sympathies to all members of the same nation, though personally unknown to him. This point once reached, there is only an artificial barrier to prevent his sympathies from extending to the men of all nations and races."[41]

Cooperation remains an unfulfilled instinct, however, unless the individuals involved communicate. Darwin argued that language spreads collective intelligence in such a compelling way that every member of a given community becomes capable of learning how to act for the common good.[42] Communication engenders reflexivity—the ability to reflect on one's own role in the construction of the social world in which one is enmeshed.

The social power of language has been multiplied many times over by the technologies that have been developed since Darwin's time. Widening the circulation and exchange of symbolic forms offers the best chance to combat ignorance and break down differences between strangers. The curse of mutual incomprehension can be broken only when people come into unavoidable contact with each other.[43] The transformation of popular consciousness is a long-term project, of course, and increased exposure to new or threatening ideas is causing bumps and bruises along the way. Today's fundamentalist religious backlash against modernity, for instance, comes at a time when the technologies of communication have made the world smaller. In the same way ants create hills to adapt to changing environments, religious zealots create their own fortresses. To fight outside influence, the faithful double-down by affirming their traditions, especially in terms of how they socialize their children. Faith-based schools are popping up everywhere in England. Madrassas saturate the Muslim world. American fundamentalists preach homeschooling.

Despite the regressive tendencies, a global conversation about morality and human rights is well under way.[44] But to grow a broader sense of community, the diverse peoples of the world must believe they have mutual interests. Ethnic diversity is least divisive when everyone identifies with common values and purposes and are willing to sacrifice personal interest for the good of the group.[45] Military organizations and most religious groups work this way. Schools and businesses try to foster a sense of togetherness. Nation-states succeed when loyalty to the state supersedes racial, religious, political, class, and cultural differences. Composing a truly universal narrative based on human rights will be necessary, therefore, to generate a widespread preference for cultural tolerance over fear and hatred.[46] Doing this poses a tremendous challenge in a world fractured by strong political, cultural, and religious allegiances. Article 1 of the United Nations Universal Declaration of Human Rights represents a step in the right direction: "All human beings are born free and equal in dignity and rights. They are endowed with reason and conscience and should act towards one another in a spirit of brotherhood."

INFORMATION EQUILIBRIUM AND TRANSPARENCY

Capitalism roared into economic and cultural domination in the 1800s with unforeseen consequences. Less than a hundred years later, Adam Smith's celebrated market economy—thought by some to be capable of checking industrialists' overzealousness and abuse—proved to be unequal to the task. In a world of demanding shareholders, pitiless bosses, and underpaid laborers, the invisible hand of the market had given up any claim to social responsibility. Since then, capitalism's predatory power has been nearly unstoppable.[47]

Relational power emerges from information control; whoever knows more benefits in the exchange. Particularly in business, the ability to gather, protect, analyze, and move information around quickly is crucial. In the wake of European and North American industrialization, newly minted capitalists found ways to control access to information about the cost and availability of raw materials, market fluctuations, competitors' activities, and broad economic trends. The major players soon dominated the global economic market. Corporations were able to influence economic activity through massive commercial advertising, lobbying, financial contributions and payoffs, and backstage political maneuvering. A substantial information asymmetry emerged, favoring the largest and most aggressive competitors.

The natural world also performs like a marketplace. Exploitative behavior carried out for immediate gain takes place throughout nature even at purely physical levels. As Darwin pointed out, "Each species tries to . . . take advantage of the weaker bodily structure of others."[48] Winners and losers emerge in the skirmishes for survival. But evolutionary outcomes are never so simple. Entirely selfish behavior does not sustain long-term viability for individuals or for species. Manipulating information to exclusive advantage is bullying behavior, and bullies don't survive long in nature. Competing individuals do better when they find ways for both to win.

In human affairs, creating an unfair information imbalance ultimately damages all parties, and they know it. Compromises are struck. The continual exchange of information (including threats) during the Cold War prevented nuclear disaster for years and led to the eventual de-escalation of nuclear arms.[49] Organizations that encourage their employees to share best practices flourish. Software open sourcing and the development of self-correcting, collaborative websites like Wikipedia® make information systems useful to anyone with access. Even automobile dealerships learned to protect their reputations by offering warranties on used cars, thereby reducing the advantage of disproportional information about the vehicle.

Information equilibrium requires a catalyst: transparency. Because it creates more openness and helps promote greater information equilibrium, transparency binds populations together against an overgrown host: dominant governing institutions. Transparency is ongoing surveillance and vigilance that makes the actions of powerful persons and institutions publicly visible and accountable.[50] From the reports of investigative journalists and the release of WikiLeaks' data and videos to the use of smartphones at protest rallies and the positioning of underwater cameras that document oil spills, transparency creates a global hall of mirrors with civilizing consequences.

Disinfecting sunlight radiates from mainstream media outlets like CNN, Channel 4, the *Guardian*, the *Washington Post*, and from alternative news sources such as the Drudge Report, *The Daily Show with Jon Stewart*, Al Jazeera, independent blogs of every political persuasion, and countless individuals using personal communication technology and social media. Powerful people and institutions are being exposed like never before. Exiting as Great Britain's prime minister, Tony Blair referred to the type of journalism that emerges from the supercharged informational landscape as a "feral beast that hunts in a pack tearing people and reputations to bits." Politicians, corporate executives, sports stars, religious preachers, pop culture celebrities, and other public figures are particularly vulnerable.[51] Corruption and abuse of every kind—including imperious foreign policies and dishonorable cultural traditions and practices—can also be revealed.[52]

COMMUNICATION POWER

Consistent with what happens generally in the animal world, communications theorist David Berlo's classic definition of effective human communication is based on the power of persuasion—the ability of a message sender to elicit the desired response from the intended receiver(s).[53] Sociologist Manuel Castells argues that today, "power is exercised by the construction of meaning in the human mind through processes of communication enacted in the global/local multimedia networks of mass communication," including the influence of individuals on the Internet.[54] In *The Future of Power*,

Joseph Nye brought these ideas together by arguing that power in the globalized world is the ability to get the outcome you want by telling the most persuasive story.[55] That's not easy to do. The social impact of any communication system—source, channel, message, receiver—is undetermined. No institutional source ever articulates but one unified ideology. The technologies that make up communications channels can never be completely controlled by political states or by corporations. The polysemous nature of messages makes them inherently open to multiple interpretations, including resistant responses. Audiences interpret and use mediated messages according to their own needs and interests, even in highly controlled environments.[56]

Moreover, the monopolistic mass-media structure that dominated the previous era is being transformed today by the diversity of contemporary media and by the enthusiastic way people everywhere embrace personal communications technologies. However, distinctions made between mass media and social media should not be drawn too sharply. All media are social and potentially influential, even subversive, as a well-known incident makes clear. When an anonymous American military employee at Abu Ghraib prison during the occupation of Iraq surreptitiously snapped camera-phone photos of naked prisoners and their taunting guards, he greatly challenged the US government's claim of moral righteousness during the Iraq War. The individual sent the images via the Internet to a friend in the United States, who released them to a commercial television station. The television station transmitted the sensational photos to its local market. The images were picked up and circulated by news outlets and Internet users around the world. Practically everyone found themselves transported inside the prison where they witnessed the abuse. Moral judgments were rendered. Global distaste for the war intensified. Public opinion in the United States took a sharp turn against the war and the Bush administration. Barack Obama emerged as a political superstar with a promise of change. The opposition party won the presidential election and took control of both houses of Congress.

The popular uprising that broke out following the 2009 presidential election in Iran was ignited by the same mixture of human initiative, technological capability, media interactivity, and symbolic complexity. Antigovernment activists sent messages that were circulated by personal communications technology, social media, other Internet sites, and international news channels. The Iranian government tried to snuff out the rebellion by shutting down mobile phone systems, by kicking foreign journalists off the streets, by censoring websites, and by strictly controlling the state-run media. But the protesters would not be deterred. Proxy servers were utilized to get around the blockage. Posts on Wikipedia, Facebook®, and Twitter® proved impossible to stop. Just as the surreptitious copying and circulating of audiocassettes had fueled the Islamic uprising that deposed Mohammad Reza Shah Pahlavi thirty years before, the latest generation of consumer communications technology facilitated popular resistance to the current regime. Encouraged by the courageous uprising in Iran, and schooled on the technologies needed for mobilizing popular support, the demand for change swept across the Middle East and North Africa two years later.

Foreign and alternative communications media, even classical literature, have long been resisted by authorities in the Arab-Muslim world in order to protect religious ideology and maintain personal privilege.[57] But the recent trends are irreversible—and global. Change is constant, and communication makes it happen. The new media landscape fosters a cosmopolitan mindset that helped provoke not only the Iranian rebellion and the Arab Spring but also the collapse of Communist regimes in the Soviet Union and Eastern Europe and, in an earlier iteration, led to the downfall of military dictatorships in Latin America.

Media fare available in the Middle East, North Africa, and south Asia began to change long before the Arab Spring erupted, helping to raise the consciousness that made the political insurrection possible. Political democratization, cultural modernization, human rights and freedoms—especially concerning the plight of women—appear as topics for discussion and debate now on many electronic media outlets in the region. Lifestyles projected on popular television programs encourage lower birthrates, more gender equality, greater tolerance for diversity, and the value of education.[58] The young adult population has developed an increasing appetite for global music videos, movies, and sports. Over twenty million people now log on to the Internet every day—more than a fourfold increase over the past six years.[59] Mass media, the culture industries, the Internet, and personal communications technology interact in ways that are profoundly changing everyday life in the region. People are warily learning how to live in greater harmony with difference. What form the political and cultural transformations will eventually take and where they will lead cannot be predicted, but that, too, will materialize under intense scrutiny.

CHANGE FROM BELOW: AN EVOLUTIONARY PRINCIPLE

Presidential candidate Barack Obama gazed out at more than eighteen thousand diverse individuals who had gathered in the athletic field house in Madison on the University of Wisconsin campus following an impressive set of victories during the primary elections in 2008. "This is what change looks like," he said, "when it happens from the bottom up." One week later, he climbed back onto the big stage, this time in Houston, Texas, to thank followers after another primary election victory. At the very moment Obama took the stage in Houston, Hillary Clinton was spinning her string of stinging defeats on television. All the news networks were carrying her concession speech live to a national audience. The public appearances of the two candidates unpredictably overlapped, causing a media crisis. Television news directors tried desperately to cover all the action by accommodating both candidates. They called for split screens—Clinton on the right, Obama on the left, or vice versa. Audio stayed with Clinton, while the other screen showed Obama bouncing up the stairs, shaking hands with locals, and surveying the joyful scene as he prepared to give another victory speech.

Ensconced in their darkened booths, TV directors were faced with a monumental decision. Do they keep the live shot of Hillary Clinton up until she finishes her speech, following the usual protocol for such a powerful person? Or do they knock her off in favor of Barack Obama when he begins to talk? When the crucial moment arrived, Hillary disappeared mid-sentence. Obama's smiling countenance prevailed, and his voice rang out. Obama was proving to be the superior candidate. He certainly was better television.

Not everyone has been pleased with how Obama is performing politically as president, and no one knows what his political legacy will be. But one thing is clear: Obama's fast political ascent stemmed from his ability to communicate persuasively. Just as biological organisms must adapt to shifting environmental conditions to survive, Obama's climb was fueled by his ability to attract an unprecedented number of small campaign contributions and to appeal to a wide range of voters, many of whom had never participated in politics before. The Internet clearly contributed to his success. At the pinnacle of the primary season, Obama's website was attracting twice the traffic of Clinton's site and five times that of Republican rival John McCain. Nearly 90 percent of the money raised in Obama's campaign came from small, online contributions. Tweets and text messages reminded voters to go to the polls. The high-tech nature of the Obama campaign created a cultural buzz and received unprecedented political results. After winning the election, Obama tried to maintain the grassroots connection with web pages like We the People, where citizens could petition the government to take action on issues that concerned them.

The Internet-driven, socially connected, bottom-up strategy that helped Obama get elected mirrors the way all of nature reaches higher plateaus. Moving from dumb simplicity to intelligent complexity, natural selection gradually produces complex instincts through the slow accumulation of slight but profitable variations. Each adaptation reflects an optimal beacon of hope. For humans, the evolutionary process reaches the conscious level and reflects our positive natural inclination toward open communication. Accumulated technological richness leads to information abundance that creates cultural options. Mass-media production and consumption have increased greatly in almost every part of the world. Intercultural interaction has intensified beyond anything we've known before. Computer and cell phone usage continues to escalate. Texting and sharing photos and videos have become universal behaviors.[60] Social networking in all its forms continues to drive a rapid increase in global connectivity, especially among young people.[61]

Powered by diverse social media, wireless connectivity, mobile broadband, cloud computing, blogging and self-publishing service providers, and Internet-enabled smart phones, the technology revolution of the Communication Age has ascended to a new level. The payoffs are tangible. The combination of advanced computer power and greater social affordance makes it possible for anyone with access to the system to know and analyze things with a level of expertise that no single human being or group could have mustered before now.[62] Diverse individuals have learned to cooperate in order to innovate, to develop a project, and to solve problems through open sourcing on the Internet. Institutions transform themselves productively by adapting smartly to the new realities. For instance, in the wake of a global economic crisis that hit the country especially hard, the government of Iceland began drafting its new national constitution by using the Internet to gather input directly from citizens. German political activists created a blog, *LiquidFeedback.org*, to inject public opinion directly into governmental decision making.

The untold possibilities that are inherent in today's communications activity are certainly not all positive. The human imagination runs in every possible direction, some of them very destructive. The bottom-up principle applies here, too. Historian Niall Ferguson writes, for example, that while communications technology helps promote political and cultural democracy overall, it has also been appropriated by "Islamists who understand how to post fatwas on Facebook, email the holy Koran, and tweet the call to jihad."[63] Norwegian Christian terrorist Anders Breivik railed against immigrants for years on the Internet, used cell phone technology to detonate the bomb in Oslo, and made calls on his mobile phone to distract the police as the mayhem progressed. Using social networking sites and secure smartphones to organize social protest can happen anywhere, as Britain's prime minister David Cameron found out in 2011. His threat to suspend protesters' access to social media sites and cell phone providers has become a familiar response to public dissent. Cyber censorship can also be applied preventively, as Chinese political authorities have demonstrated for years.

Since the invention of the integrated circuit (also known as the silicon chip)—the foremost technological advance of the late twentieth century—most high-tech trends run in a positive direction. They are designed to produce enhanced forms of social intelligence. Nearly two hundred years after the Industrial Revolution rocked Great Britain and the rest of the world, evolutionary concepts like the "innovation economy," "open development," "invention capital," and the "ecology of innovation" came to define the current era. The Communication Age represents a milestone in evolutionary history. As Spencer Wells, director of the National Geographic Society's Genographic Project, says, "Everywhere there is a feeling that the world is in flux, that we are on the brink of a historic transition, and that the world will be fundamentally changed somehow in the next few generations."[64]

THE PROMISE OF HUMAN EVOLUTION

Natural history is the story of ceaseless change. For humans, change generally means improvement. Medical science makes it possible to increase or eliminate sexual fertility, defeat diseases, prolong the life span, and share vital organs. Modern transportation overcomes physical limits on personal mobility. Natural resources are turned into energy sources. DNA and the human genome unlock inner mysteries, while satellites and space stations fly overhead.

Having spent a lifetime studying the morphological characteristics of living things, Charles Darwin recognized that while nothing stays the same for long, an elemental "unity of type" also persists in nature. Organic beings of the same class exhibit a "fundamental agreement in structure," Darwin said, "which is quite independent of their habits of life." He called the reproduction of these structural similarities over time the "unity of descent."[65] Long-term cultural histories have forged differences among human populations, but strong physical and behavioral uniformities across human groups also exist. This fact contributed greatly to Darwin's key conclusion that we all descend "from a single parent-stock, which must have been almost completely human in structure, and to a large extent in mind, before the period at which the races diverged from each other."[66]

How, then, do we finally make sense of ourselves as diverse communities of descent that grew from a common ancestry? How can the best way forward for humanity be achieved? In the biological world, Darwin noted, habits generally change first and structure, afterward.[67] Nonhuman organisms usually change their habits out of necessity and without conscious awareness. Humans have the advantage of being able to shape their behavior by using keen foresight, rationality, creative thinking, and moral determination.

Shared knowledge and new habits come into being through repeated communicative exchange. Over time, an inherent synergy between human communication and cooperation generates ideas, trust, and a sense that we can willfully exercise control over our shared destiny. New modes of community are created and re-created endlessly. Political, economic, and cultural development become tightly interrelated. Modern values disseminate as the global middle class expands.[68] Cultural development proceeds from the bottom up even at the global level, but enlightened leadership from political and cultural entities worldwide will also be essential.

This vision is not unachievable. Genetic predispositions and cultural traditions do not determine future developments. No other species can subordinate its genetic interests to noble concerns or change the world by conscious decision and sheer determination. Positive thinking alone has great adaptive value. As Richard Dawkins puts it, "We have the power to defy the selfish genes of our birth and, if necessary, the selfish memes of our indoctrination. We can even discuss ways of deliberately cultivating and nurturing pure, disinterested altruism—something that has no place in nature, something that has never existed before in the whole history of the world."[69] Acting charitably does not have to be motivated for evolutionary reasons centered solely on immediate self-interest; it can also take the form of enlightened decision making that overcomes the tit-for-tat nature of reciprocal altruism. That's evolutionary, too.

Writing clearly and with a tender heart, Charles Darwin explained in his books how evolutionary currents flow from the simple to the complex while never completely leaving their origins. Each one of us carries something of every other living thing as part of our biological makeup. All living things descend from the same seed; the potential for all humankind to live harmoniously is written into our DNA. Yet social injustice and human conflict endure.

As he departed Brazil's northeast shores for the last time aboard the *Beagle* on August 19, 1836, Darwin vowed he would never visit a slave country again. He was haunted by the screams of African slaves being beaten and tortured—a little boy horsewhipped on the head by his master for serving Darwin a "not quite clean" glass of water; the fingers of female slaves crushed by their owner, an old woman in Rio de Janeiro. Darwin was sickened by how the slaves' dignity had been destroyed and their families ripped apart in colonies throughout the New World. In *The Voyage of the Beagle*, published three years after Darwin returned from his epic journey, he criticized his fellow Britons and their American descendants for uttering a "boastful cry of liberty" while at the same time enslaving their fellow humans.[70] To Darwin's great relief, the British and American slave trade and ownership ended in his lifetime.

Charles Darwin's cultural experiences abroad and at home were just as important to him as his finely detailed studies of animal and plant life in figuring out what to finally conclude about the descent of humans. In bringing together the seemingly incompatible ideas of diversity and unity in a comprehensive way, one overriding inference can be drawn from Darwin's magnificent theory: evolution is not determined and it is not random. It is precisely in that unmapped space between determination and randomness where our special talents as communicators will decide the future.

GLOSSARY

Abrahamic religions. The three dominant monotheistic religions: **Judaism, Christianity,** and **Islam.**

adaptation. (1) Characteristic or ability of a living organism to adjust to environmental conditions. (2) Characteristic or trace of individual organisms or species that allows survival in a specific environment.

agnostic. Person who professes any philosophical doctrine or viewpoint that deems it impossible to know God or gods and at the same time suspends judgment about the existence of any divinity.

altruism. (1) Term coined by nineteenth-century French philosopher Auguste Comte that refers to the human instinct to care for other human beings. According to Comte, altruistic instincts can be improved by appropriate education. (2) Selfless love for the other; abnegation.

analogical. Relation of similar continuity between representations, facts, or things; the opposite of **digital.**

animal communication. Under the form of a behavior or a movement, a message sent by an animal that affects, changes, or influences another animal behavior.

antithesis. Formal principle rigidly contrasting two messages with conflicting meanings. Darwin recognized the principle of antithesis in all animal expression: if a dog perks up its tail to indicate a willingness to fight, the opposite message will take the opposite form: the dog cowers and hides its tail between the legs.

apostasy. Renouncement or abandonment of a religion or a belief.

artificial selection. Production of a biological variety of animal or plant through human action; the opposite of **natural selection.**

atheist. One who doesn't believe that God or gods exist or can be proven to exist.

belief. (1) Deep and personal conviction about anything, any idea, or any person. (2) Attitude or certainty about something (an ideology, a person, a religion, and so on) that is deemed to be the path to redemption and salvation. (3) Sense of truth, often without rational reasons.

bipedalism. (1) Specific capacity of some animals to move using two legs and feet. (2) Human capacity to use feet for moving in an environment.

Blombos Cave. Archaeological site located on the Southern Cape Coast of South Africa, where the earliest evidence of art and jewelry making was found.

cell. Structural, functional, and microscopic biological entity made of cytoplasm, plasmatic membrane, and genetic material.

Christianity. The set of religious interpretations based on Jesus Christ's ethics and its promise of redemption, comprising hundreds of theological variations ranging from Catholicism to Evangelical Protestantism and other modes of Christian orthodoxy.

code. System of shared rules by senders and receivers of messages that ensures the possibility of communication.

cognition. (1) Act or effect of knowing. (2) Faculty or process that enables the acquisition of knowledge.

collaboration. Act or effect of participating in work or labor with other individuals; cooperation.

common descent with modification. Darwin's formulation of evolutionary theory's foundational principle. States that all organisms spring from a common ancestor; explains the biological variety and diversity in organic life.

communication. Process involving transmission and reception of messages between source and destination, whether through natural and physical resources or technical devices.

communities of descent. Darwin's concepts used to describe ethnic and cultural groups; analogous to the notion of common descent with modification.

competition. In ecological settings, the interaction of individuals of the same or different species; in an environment of scarce or limited resources, the growth and survival of one individual or population affecting others negatively.

complexity. Quality of what is complete or whole with a greater or lesser degree of coherence and articulation, in which each component interacts with other parts of the totality in a relationship of interdependency or subordination.

connectivity. Capacity or possibility of a computer, informational device, or program to operate in social or technical networks.

consciousness. (1) Feeling or knowledge that allows self-awareness; the experience, understanding, and grasping of one's internal world. (2) The essence or totality of attitudes and opinions, or worldview, held by an individual or group.

contingency of nature. Condition of perpetual uncertainty indicating that there is no plan or design in nature.

cooperation. Act or effect of collaborating with another being. In a population of intrinsically self-centered organisms set on surviving at all costs, cooperation at first became an enigma for the Darwinian theory of natural selection. Because cooperation has been proven to be evolutionarily beneficial to social species, especially primates, it has become a selected trait.

cortex. External layer of the brain's cerebral hemispheres; the seat of nervous system operations, which are translated into voluntary movements and actions; the brain's outer gray matter.

cost. Effort, work, or expense employed in the production of goods, services, and messages. In evolutionary communication, the cost of sending a signal is an important index of biological fitness: the organism can only repeatedly pay the price of emitting a message that is truly and honestly cast within the range of its possibilities; therefore, evolution favors honesty. In economics, cost is the value of merchandise calculated monetarily, taking into account the investment, time spent on its production, and margin of profit reserved for the entrepreneur.

creationism. Religious and moral doctrine based on the biblical narrative of the book of Genesis, which states that the organic world was created from nothing in a perfect and unchanging manner as an act of God's will.

cultural development. Process constantly acting on human societies that transforms and generates new cultural forms in the image of biological evolution.

cultural drift. Notions, patterns, representations, and cultural variations resulting from chance, not from selection.

cultural imperialism. Dominance of one human group over another through the spread and the invasive dissemination of cultural products to the detriment of cultural and social diversity.

culture. System or set of patterned behaviors, beliefs, knowledge, and practices (ritualistic, economic, religious, and communicative) that distinguishes and differentiates human groups.

digital. Relation of discontinuous and binary opposition between representations, facts, or things; the opposite of **analogical.**

dimorphism. Property of two different types of organisms that are members of the same species. See also **sexual dimorphism.**

diversity. In ecological terms, index and ratio that considers the number of species and their abundance in an area, community, or sample. Diversity indicates biological invention created by **natural selection**; in human evolution, diversity can be measured in terms of artifactual invention.

DNA. Deoxyribonucleic acid. Nucleic acid whose sequencing generates genetic information for living beings and viruses.

double helix. Structural form of the double-stranded molecules of **DNA.**

ecological niche. Position and location of an organism in a specific biological community.

essence. In standard philosophical vocabulary, the most basic, central, and complete characteristic or set of traits determining definitively what shapes existence as a whole.

essentialism. Any speculative doctrine that sustains, argues, and advocates the primacy of an essence capable of determining the actual existence of a being.

evolution. (1) Process through which organisms and species mutate over the course of time. (2) Theory according to which organisms and species modify their characteristics as a result of biological mutation and **natural selection.**

expression. Act or effect of manifesting oneself through words, gestures, artistic works, dance, facial displays, and so on.

faith. In the Christian tradition, one of the three cardinal virtues through which humanity accepts God's revelation made by the Church. According to Scholastic theology, faith doesn't require rational arguments but merely the obedient embrace and acceptance of unquestionable assumptions, even if they seem irrational and absurd.

falsifiability. Intrinsic possibility of a real scientific theory that, as a result of testing, can be totally or partially refuted. Falsification is the demarcation line of science and nonscience as to what makes scientific investigation a progressive, self-correcting enterprise.

feedback. (1) Retroactive reaction to stimuli. (2) Information that the sender of a message obtains from the reaction of the receiver. Feedback information is useful to evaluate the effectiveness of a transmitted signal.

***FOXP2* gene.** Protein located on human chromosome 7, whose mutations can cause speech and linguistic disorders, thus leading to the hypothesis that it may be a genetic condition for human language.

gene. The fundamental unity of heredity composed of a segment of a string of **DNA** capable of synthesizing a protein. The dominant strategy of any gene is to replicate itself at all costs. For this reason, Richard Dawkins coined the term *selfish gene.*

genetic drift. Heritable organic variation resulting from chance, not from selection.

genetic relatedness. Degree of kinship between organisms.

genome. (1) Complete set of **genes** in a living being. (2) Set of all **DNA** molecules present in a cell of a living being.

globalization. (1) Worldwide integration resulting from cultural and economic exchange, made possible by new information and transportation technologies as well as the neocolonialist actions of transnational corporations. (2) The flow of people, symbolic forms, commodities, money, information, and ideas on a global scale, which some critics argue is creating a homogenous world culture.

group selection. An explanation proposed by ornithologist V. C. Wynne-Edwards for the development of altruistic behavior among animals and humans based on the assumption that the group, not the individual (as in **natural selection**), is the main and dominant beneficiary of a specific behavior. Group selection takes natural selection to a collective level. See also **kin selection.**

hegemony. (1) Process through which dominant ideology is transmitted, consciousness is formed, and social power is exercised. (2) The power or dominance one social group holds over others. Rather than direct manipulation of people against their interests by violence or coercion, hegemony depends on social actors accepting their subordinate status as normal. Ideology-dispensing institutions such as governments, religious organizations, schools, corporations, and mass media reinforce each other by perpetuating the status quo as common sense.

hidden ovulation. A type of ovulation (common among humans) that displays no evident or visible sign. See also **ovulation.**

hominid. Family of **primates** consisting of humans and their fossil ancestors.

Homo neanderthalensis. Family of hominids that lived side by side with *Homo sapiens* and went extinct. Their fossil record was found in Neandertal, an area of Germany.

Homo sapiens. Species of **primate** to which the modern human belongs.

hybridity. (1) Absence of regularity, constancy, or stability generated by the crossing of biological and cultural variants, lineages, or species. (2) The fusing of symbolic forms often facilitated by the creative mixing of media imagery.

ideology. (1) Any set of philosophical, political, cultural, or social ideas held by an individual or group. (2) System of ideas sustained and adopted by individuals or social segments and expressed in communication for the purpose of rationalizing and defending moral, religious, political, or economic self-interest or social commitments.

inclusive fitness. In W. D. Hamilton's theory, the type of fitness that favors the reproductive success of relatives, even at personal cost. Inclusive fitness is directly proportional to genetic relatedness.

index. A type of sign created by the material relationship of cause and effect. Weather vanes are indices because the pointers of the vane are moved in a certain direction as a result of the wind; the condensation that forms on the external surface of a glass of cold liquid is another index.

Industrial Revolution. Historical era of staggering social and economic transformation that began around 1760 in England and spread to other countries around the world. More than the mere application of mechanical and steam power to economic production, the Industrial Revolution transformed of the whole fabric of social life. With it came a revolution in customs and fundamental changes in the structure of family life, followed by an unprecedented urban explosion in population. In *The Manifesto of the Communist Party*, Karl Marx and Friedrich Engels summarized the experience in a celebrated sentence: *all that is solid melts into air*.

information. (1) Transmission or reception of a statement, utterance, or assessment. (2) Knowledge or fact of general and common interest transmitted by vehicles of mass communication. (3) Message transmitted by new technologies of communication.

information equilibrium. Communication strategy of cooperative interaction. In the case of an interaction where both players want to win, cooperation can emerge through the reduction of any information asymmetry that exists between them.

information technology. Technological and computational equipment that facilitates the storage, transformation, dissemination, reception, and optimization of information circulating in social networks.

intelligent design. A contemporary version of **creationism** whose argument claims, in a way similar to **natural theology**, that the complexity of the natural word requires the existence and action of a complex creator.

intention. What is aimed to be reached, consciously or unconsciously.

interactor. What or who interacts with other biological or social agents.

Islam. Monotheistic religion created by the Arab prophet Muhammad that blends religious faith with sociopolitical organization. Islamic principles, tenets, and prescriptions are codified in the Koran.

Judaism. (1) Monotheistic religion of Jewish people, tracing its ethnic and spiritual roots to Abraham. The faith's ideological principles are expressed in the Bible and the Talmud. (2) Jewish culture and civilization.

kin selection. A level of evolutionary selection proposed by theoretical biologist W. D. Hamilton that emphasizes kinship relations over the absolute interests of the individual or group. Kin selection favors relatives even at the expense of personal survival or reproduction; the concept of kin selection is considered by many to be a powerful explanation of the behavior of neuters in insect colonies. **inclusive fitness** is an effect of kin selection. See also **group selection**.

language. Any systematic or nonsystematic way of conveying ideas, feelings, or thoughts, and for producing interaction by employing conventional or nonconventional signs.

life. (1) Period in the existence of a living organism between the extreme points of birth and death. (2) A system or population submitted to the principle of evolutionary **natural selection** marked by heredity, **mutation**, and the spreading of replicated mutation.

magical thinking. Act or effect of using mental powers to transform the natural world through spells, enchantment, and rituals.

Marxism. Any doctrine or social theory deriving from the writings of Karl Marx or Friedrich Engels. Regardless of how different and conflicting the several currents of Marxism may appear as a continent of ideas, they all share a common view: an emphasis on the economic conflicts and contradictions of social life.

meaning. What something signifies or represents to an individual. Meaning is not inherent in material or symbolic forms but is constructed by those who interpret the objects, experiences, or symbols according to their own orientations, interests, and competencies.

media. (1) Any system that allows the diffusion of information and transmission of messages. (2) Communications technologies capable of sending messages over broad expanses of space and time.

meme. Any cultural form, theme, trait, or fragment that travels from individual to individual in a human group. The concept of *meme* unifies biological replication and cultural transmission. See also **memetics**.

memeplex. Association or aggregation of **memes**.

memetics. The study of the actions and processes of **memes**. See also **meme**.

mimicry. Relationship of similarity between organisms that gives the imitating one an advantage in the process of **natural selection**.

mirror neuron. A specific type of neuron that fires when the organism does a specific action or when it observes another organism performing the same act.

modernity. Historical period that developed after the dissemination of ideas originating during the Enlightenment, wherein humanity came to be conceived as autonomous, self-sufficient, universal, and ruled mainly by reason.

Mormonism. Apocalyptic and messianic Christian doctrine created by Joseph Smith Jr. in the United States. Smith compiled and promoted the Book of Mormon, upon which the doctrine is predicated.

morphology. Form and ordering of the parts of an organism, identified by direct observation.

multilevel selection. Geneticist George R. Price's contention that **natural selection** can work simultaneously and progressively at individual and group levels.

mutation. (1) Act or effect of altering, changing, or transforming. (2) Sudden alteration of the genetic makeup of an individual without direct connection to its immediate ancestors but capable of being transmitted to its descendants.

natural law. Set of material principles constantly at work in the natural world.

natural selection. Survival and/or differential reproduction in a population of living organisms as a result of being better adapted to the environment and to the detriment of the less fit biological forms.

natural theology. A philosophical approach to religion that uses common observation of nature as an argument in favor of God's existence. Natural theology doesn't rely on the authority of a presumably sacred text because it contends that through contemplation one can see nature's perfect design, which can only lead to the certainty that a designer must have created the world where one lives. The assumptions of natural theology go back to Thomas Aquinas in his *Summa Theologica*, but they were immensely influential in Darwin's day. **Intelligent design** is a contemporary version of natural theology.

noise. (1) Any interference or disturbance causing loss of information during the transmission of a message. (2) In telecommunication, an undesired or unintended electric signal that interferes in the transmission of a message.

organism. (1) Individual form of life manifested as a plant, animal, fungus, and so on. Organisms are structured by organs and other features that perform vital functions. (2) The organic constitution.

ovulation. Act or effect of liberating an egg from the ovaries with the possible consequence of it being fertilized. See also **hidden ovulation.**

Pascal's wager. Argument put forth by French mathematician and theologian Blaise Pascal. According to Pascal, the adoption of religious faith amounts to a bet on the existence or nonexistence of God. If God doesn't exist, the person who bets loses nothing, but in the case of the existence of God, the loss and punishment are great. It is therefore rational to place a wager in favor of the existence of God. Pascal reduces the question of religious choice to a game of simple probabilities.

personal cultural programming. A user-centered process opened up by the growth of media, information technology, and the wide circulation of cultural resources that grants individuals increased creative power over conventional culture to satisfy personal needs. Change in the locus of cultural power from institutions to individuals. See also **superculture.**

phoneme. The sonic unity of speech, deemed to be the minimal distinctive feature of a language, organized in a linguistic system by means of opposite sound values without which words with different meanings could not be identified. Phonemes have no meaning in themselves; their semantic value depends on their structural and systemic organization. Phonemes shouldn't be confused with letters because they aren't graphic but acoustic images.

polysemy. Multiplicity of possible meanings in a word or set of words.

predation. Act or effect of capturing prey with the purpose of feeding.

predator. Animal that hunts with the purpose of feeding itself.

prey. Animal that is captured or hunted by a predator that has the goal of feeding itself.

primate. Species of mammals comprising humans, apes, lemurs, monkeys, and related living forms endowed with large brains, binocular and well-developed vision, and hands and feet with five digits.

primeval soup. A hypothesis postulated by Alexander Oparin in 1924. Oparin's theory springs from the recognition that an atmosphere with oxygen doesn't favor the synthesis of organic compounds for the generation of life; however, that doesn't mean that there can't be spontaneous generation if some conditions are met in an atmosphere without oxygen but under the action of sunlight. At about the same time, British biologist J. B. S. Haldane formulated a similar contention: that life came from prebiotic oceans. The later vindication of both assumptions delivers a hard blow to creationists' speculations.

racism. (1) Set of theories, doctrines, or opinions that establishes a hierarchy between ethnicities. (2) Derogatory or demeaning opinion about individuals or groups of another race.

reciprocal altruism. A kind of altruistic behavior among living organisms performed with the assumption that reciprocation will occur in the future. The creator of the concept, evolutionary biologist Robert Trivers, identified reciprocal altruism as a type of altruism that doesn't surrender self-interest.

redundancy. Quality of what is transmitted repetitively with the intention of ensuring that the message is adequately received; redundancy is a recurrent feature of communication systems.

reflexivity. Any quality or attribute revealed in the relation of an element with itself. To communicate is not reflexive, but the act of the communicator understanding the rules of communication and his or her role in the way meaning is derived from communicative exchange is an effect of reflexivity.

religious altruism. A type of **altruism** based on religious doctrines.

religious communication. Act or effect of communicating ideas that are derived directly or indirectly from religious tenets or principles.

religious exceptionalism. Sense of individual and collective uniqueness (and often superiority) produced by the adoption of a religious set of beliefs.

religious hegemony. Act or effect of exercising social and political power through consent that is fabricated by the intentional persuasive use of religious principles, content, and rituals.

religious socialization. Act or effect of developing individuals for life in a community through repeated religious indoctrination.

representational art. Art that depicts with greater or lesser degree of mimetic faithfulness the objects, scenarios, and figures of the natural world.

Scientology. Religion and pseudoscientific movement created in the 1950s in the United States based on Dianetics—a set of ideas and practices regarding the metaphysical relationship between the mind and body, invented by a science fiction writer. Scientology argues for emphasizing the importance of the soul, spirit, and physical world.

selected trait. Biological characteristic or capacity singled out by natural selection. Selected traits spread incrementally in organic populations.

selection pressure. Force triggered by the process of natural selection in organic life, leading to an advantageous outcome for some living forms to the detriment of others.

selfish gene. See **gene.**

semiotics. The study of the constitution and mechanisms of representative signs.

sex. (1) The splitting of sexual species into males and females according to different organic traits and reproductive roles. (2) Among animals, a set of characteristics that distinguishes males and females.

sexism. Discriminatory behaviors, attitudes, or opinions based on sexual characteristics.

sexual dimorphism. Group of visible differences (form, size, color, and so on) recognizable in masculine and feminine members of the same species. See also **dimorphism.**

sexual selection. In Darwin's evolutionary theory, an important mode of biological selection resulting not in the elimination of competitors but in the generation of offspring: the cause of biological differences and competitive advantages among members of the same species.

shaman. Individual who, through utterances, physical movements, and specific rituals, demonstrates and manifests medical, magical, and divinatory powers.

sign. The representation of something to someone under a particular circumstance. As long as a material element performs the function of representation, it can be deemed a sign. The concept of *sign* includes verbal and nonverbal, cultural and natural, codified and non-codified modes of expression.

social media. Media designed for social interaction using web-based technology as opposed to industrial, centralized, and traditional media. Bottom-up, not top-down, circulation is a definitive trait of social media.

sociobiology. Comparative study of the genetic basis, evolutionary history, and biological foundations of social behavior in animals and humans.

speciation. Evolutionary process by virtue of which organic variation and new species come to life; being both natural, if occurring in the wild, or artificial, when produced by husbandry or generated in a laboratory.

species. In a hierarchical biological classification, a taxonomic category below the category of genus. The identification of species is based on the recognition of similar morphological traits among individuals and their parents; members of the same species breed among themselves and generate fertile descendents.

superculture. Personalized selection of symbolic and material resources made by an individual to create a constantly modifiable cultural profile or identity. See also **personal cultural programming.**

superorganism. In ecology, a community of individual organisms that collectively creates an organized entity (hives or mounds, for example) that allows the organisms to survive and reproduce.

survival of the fittest. Expression coined by biologist and philosopher Herbert Spencer after reading Darwin's *On the Origin of Species*, referring to survival as a fitness indicator of the individual organism. The expression has been a source of criticism of Darwinian theory and of antiquated Social Darwinism. In the fifth edition of *On the Origin of Species*, Darwin equated survival of the fittest with **natural selection**. However, modern biology doesn't reduce natural selection to mere survival because natural selection should always imply the differential reproduction of organisms.

symbol. A type of sign generated by the adoption of common and shared rules of formation. Symbols are fundamentally cultural as opposed to natural indices. A word is a symbol since it isn't directly motivated by the object it represents, and it springs from an arbitrary system of rules. The Christian cross is a symbol: a set of conventions create this particular representation without taking into account its origin as a torture device.

symbolic creativity. Human capacity to transform, combine, and generate new forms of representations that have impact over other members of the community; the use of symbolic forms, especially media imagery, to influence the course of social action and events.

time. Relative duration of things that creates in the human mind the sense of past, present, and future. Evolutionary time doesn't follow the image of a straight and unidirectional flow; the Darwinian notion of *common descent with modification* implies that origin is present in descendants. If one looks at the point of view of the future, life-forms branch out in endless variations; looking at the past, the variations converge to a common point of origin.

transmutation. Same as **mutation**; the term *transmutation* was current in the biological vocabulary of Darwin's time. Darwin substituted it for *evolution* because the effects of transmutation are evolutionary changes.

universal grammar. In Noam Chomsky's linguistic theory, the faculty of acquiring language or an initial state of mind (part of humanity's biological inheritance) that, under standard conditions of social interaction, allows the acquisition and internalization of the language to which an infant is exposed.

world literature. In *The Manifesto of the Communist Party*, Karl Marx and Friedrich Engels refer to the phenomenon of a universal literature, a sign of their premonition of the future emergence of global modes of communication.

NOTES

INTRODUCTION

1. Electronic Arts' *Spore*™—another interactive game allowing players to control the development of a species—may also help popularize evolutionary thinking among video game players.

2. Stuart A. Kauffman, *Investigations* (New York: Oxford University Press, 2000). The "adjacent possible" is discussed by Steven Johnson in *Where Good Ideas Come From* (New York: Riverhead Books, 2010).

3. A similar phenomenon occurred during the student-worker uprising in China in 1989. While resistance shown by workers within the state media system didn't lead to immediate revolutionary change, it helped pressure the government to undertake major reforms in the following years. See James Lull, *China Turned On: Television, Reform, and Resistance* (London: Routledge, 1991).

4. H. Paul Grice, "Meaning," *Philosophical Review* 64 (1957): 377–88; H. Paul Grice, "Logic and Conversation," in *Syntax and Semantics, Vol. 3: Speech Acts*, edited by P. Cole and J. Morgan (New York: Academic Press, 1975), pp. 43–58; Michael Tomasello, *Origins of Human Communication* (Cambridge, MA: MIT Press, 2008).

5. Nicholas Wade, *The Faith Instinct* (New York: Penguin, 2009), p. 7.

6. Nicholas Wade, *Before the Dawn* (New York: Penguin, 2006), p 34.

7. Carl Zimmer, "Crunching the Data for the Tree of Life," *New York Times*, February 10, 2009.

8. Manuel Castells, *Communication Power* (New York: Oxford University Press, 2009), p. 427.

9. John Hawks et al., "Recent Acceleration of Human Adaptive Evolution," *Proceedings of the National Academy of Sciences* 104 (2007): 20753–58.

10. Sherry Turkle, *Alone Together* (New York: Basic Books, 2011).

11. Wallace Stegner, *This Is Dinosaur* (Boulder, CO: Robert Rinehart, 1985), p. 17.

CHAPTER 1. THE GREAT CHAIN OF COMMUNICATION

1. The chain metaphor became so popular and useful it was later elevated to "The Great Chain of Being," the name of a series of lectures given by American philosopher Arthur Lovejoy at Harvard University in the 1930s. It was also the title of his classic volume describing how the history of ideas in the Western world has been imagined throughout the centuries: Arthur Lovejoy, *The Great Chain of Being* (Cambridge, MA: Harvard University Press, 1936). Almost fifty years later, Ernst Mayr used the chain metaphor to describe a biological discovery made by natural theologians, the tradition from which Darwin broke away. Mayr wrote, "The discovery of infusorians and zoophytes seemed to confirm the Great Chain of Being leading to man." See Ernst Mayr, *The Growth of Biological Thought: Diversity, Evolution, and Inheritance* (Cambridge, MA: Belknap, 1982), p. 104. Infusorians are a class of protozoa, unicellular organisms. Zoophytes are small animals like sea anemones and sponges.

2. Charles Darwin, *The Descent of Man* (Princeton, NJ: Princeton University Press, 1981), p. 602.

3. Charles Darwin, *The Origin of Species*, 2nd ed. (New York: Gramercy, 1979), pp. 160–61.

4. Ibid., p. 172.

5. Frank Newport, "On Darwin's Birthday, Only 4 in 10 Believe in Evolution," Gallup®, February 12, 2009, http://www.gallup.com/poll/114544/Darwin-Birthday-Believe-Evolution.aspx (accessed February 11, 2009); *Newsweek*, April 9, 2007, pp. 31–37.

6. Ed Stoddard, "Poll Finds More Americans Believe in Devil than Darwin," Reuters UK, November 29, 2007, http://uk.reuters.com/article/2007/11/29/us-usa-religion-beliefs-idUKN2922875820071129 (accessed August 21, 2008).

7. A comprehensive and well-supported dismantling of "intelligent design" is Jerry Coyne, *Why Evolution Is True* (New York: Penguin, 2009).

8. Darwin, *Descent of Man*, p. 643.

9. "Public Acceptance of Evolution in Science," National Center for Science Education, August, 15, 2006, http://ncse.com/news/2006/08/public-acceptance-evolution-science-00991 (accessed January 30, 2007).

10. Dennis Quammen, *The Reluctant Mr. Darwin* (New York: Norton, 2006), pp. 14–15.

11. National Center for Science Education, "Public Acceptance of Evolution in Science."

12. Ibid.

13. Darwin, *Descent of Man*, p. 629.

14. Nicholas Wade, *Before the Dawn* (New York: Penguin, 2006), p. 5.

15. Pat Robertson, comments made on Christian Broadcasting Network's *700 Club*, November 10, 2005.

16. *Kitzmiller v. Dover Area School District*, United States District Court, M.D., Pennsylvania, December 20, 2005.

17. Associated Press, "Kansas Board Boosts Evolution Education," MSNBC, February 14, 2007, http://www.msnbc.msn.com/id/17132925/ns/technology_and_science/t/kansas-board-boosts-evolution-education/#.Ty_zucU9lm0 (accessed March 15, 2007).

18. T. Trent Gegax et al., "Doubting Darwin," *Newsweek*, Daily Beast, February 6, 2005, http://www.thedailybeast.com/newsweek/2005/02/06/doubting-darwin.html (accessed February 22, 2005).

19. Todd Burpo with Lynn Vincent, *Heaven Is for Real: A Little Boy's Astounding Story of His Trip to Heaven and Back* (Nashville: Thomas Nelson, 2010). The story was reconstructed by the boy's father, a minister in a small Nebraska town. Lynn Vincent is also the coauthor of Sarah Palin's autobiography *Going Rogue*.

20. Steve Bradt, "Molecular Analysis Confirms T. Rex's Evolutionary Link to Birds," *Harvard Gazette*, April 4, 2008, http://news.harvard.edu/gazette/story/2008/04/molecular-analysis-confirms-t-rexs-evolutionary-link-to-birds/ (accessed May 14, 2010); Luis M. Chiappe, *Glorified Dinosaurs* (New York: Wiley-Liss, 2007); Coyne, *Why Evolution Is True*, p. 137.

21. Coyne, *Why Evolution Is True*, p. xiii.

22. Ibid., p. 16. A "true" theory can be falsified because, unlike religious dogma, scientific claims are always open to new discoveries, evidence, and interpretation.

23. Coyne, *Why Evolution Is True*, p. 192.

24. Ibid., p. 224.

25. Stephen Hawking and Leonard Mlodinow, *The Grand Design* (New York: Bantam, 2010), p. 180.

26. There was one notable exception to this pervasive tendency. In a letter to João IV, the king of Portugal, the great Portuguese-Brazilian writer Father Antonio Vieira wrote in majestic and glittering prose to bristle against the treatment of Brazilian natives. The barbaric violence of the colonizers horrified him. Vieira asked the king to end slavery. But when João IV died, Vieira was banished from the Portuguese court and eventually exiled and charged with unorthodox beliefs by the Inquisition. Vieira kept on preaching and assailing the colonial masters. In 1691, near the end of his life, Vieira was still writing to another king, D. Pedro II, complaining of abuses against slaves. See Antonio Vieira, *Cartas* (Rio de Janeiro: W. W. Jackson, 1948). Unfortunately the enslavement of indigenous Brazilians continued, and the

slave trade expanded with large-scale importation of labor from Africa, especially from areas now known as Angola and Mozambique.

27. Adrian Desmond and James Moore, *Darwin's Sacred Cause* (New York: Houghton Mifflin Harcourt, 2009).

28. Darwin, *Descent of Man*, p. 591.

29. Howard Fineman, *The Thirteen American Arguments* (New York: Random House, 2008).

30. Thomas R. Malthus, *An Essay on Population, or A View of Its Past and Present Effects on Human Happiness* (London: John Murray, 1798).

31. *Nature* 455 (October 23, 2008): 1007.

32. Janet Browne, *Charles Darwin: A Biography, Volume 1: Voyaging* (New York: Knopf, 1995), p. 224.

33. Darwin, *Descent of Man*, p. 631.

34. Ibid., p. 156.

35. Wade, *Before the Dawn*, pp. 9, 99.

36. "Hitler Jewish? DNA Tests Show Hitler May Have Had Jewish and African Roots," *Huffington Post*, December 25, 2005, http://www.huffingtonpost.com/2010/08/25/hitler-jewish-dna-tests-s_n_693568.html, (accessed August 25, 2010).

37. Darwin, *Descent of Man*, pp. 583–86.

38. Donald E. Brown, *Human Universals* (Philadelphia: Temple University Press, 1991).

39. Steven L. Kuhn and Mary C. Stiner, "What's a Mother to Do?" *Current Anthropology* 47 (2006): 953–80; Nicholas Wade, "Equality between the Sexes: Neanderthal Women Joined Men in the Hunt," *New York Times*, December 5, 2006.

40. Claudia Goldin and Lawrence F. Katz, *The Race between Education and Technology* (Cambridge, MA: Belknap, 2010).

41. Malte Andersson, *Sexual Selection* (Princeton, NJ: Princeton University Press, 1994).

42. Alison Jolly, *Lucy's Legacy: Sex and Evolution in Human Intelligence* (Cambridge, MA: Harvard University Press, 1999).

43. Coyne, *Why Evolution Is True*, p. 160.

44. Darwin, *Descent of Man*, pp. 168–70.

45. "Is It Possible Megan Fox Is Overexposed?" *San Jose Mercury News*, September 2, 2009.

46. United Nations Development Programme, *Human Development Report: Cultural Liberty in Today's Diverse World* (New York: Oxford University Press, 2004).

47. Darwin, *Descent of Man*, p. 586.

48. Darwin, *Origin of Species*, p. 128.

49. The first comprehensive descriptions of culture as a system of life didn't arrive until Edward B. Tylor's *Primitive Culture* was published 1871, the same year Darwin published *The Descent of Man*: Edward B. Tylor, *Primitive Culture: Researches into the Development of Mythology, Philosophy, Religion, Art, and Custom* (London: John Murray, 1871); Bronisław Malinowski's classic volumes of field research appeared half a century later: Bronisław Malinowski, *The Argonauts of the Western Pacific* (London: Routledge & Kegan Paul, 1922); Bronisław Malinowski, "Culture," in *The Encyclopedia of Social Sciences*, vol. 6 (New York: Macmillan, 1931), pp. 621–46. Only then does the notion of culture become a recognized theoretical alternative to biological, racial, and environmental explanations of human behavior.

CHAPTER 2. COMMUNICATING TO SURVIVE

1. Pew Global Attitudes Project, "Widespread Support for Banning Full Islamic Veil in Western Europe," Pew Research Center, November 8, 2010, http://www.pewglobal.org/2010/07/08/widespread-support-for-banning-full-islamic-veil-in-western-europe/ (accessed November 8, 2010).

2. Charles Darwin, *The Descent of Man* (Princeton, NJ: Princeton University Press, 1981), p. 99, Charles Darwin, *The Descent of Man*, 2nd ed. (Amherst, NY: Prometheus Books, 1998), and Charles Darwin, *The Descent of Man, and Selection in Relation to Sex* (London: John Murray, 1871), p. 126. Richard Dawkins takes the point further, saying he considers forced religious socialization of the young to be a form of child abuse in *The God Delusion* (New York: Houghton Mifflin, 2006). In a television documentary, Dawkins argued, "There is no such thing as a Christian or Muslim child, only a child of Christian or Muslim parents" (Richard Dawkins, "Root of All Evil?" IWC Media documentary, Channel 4, London, 2006). In late 2007, Muslim leaders in Britain proposed a code of conduct to promote civic responsibility in the wider society and recognize women's rights. Many Muslim groups in the United Kingdom resisted this effort, however, fearing that government would wield too much power over religion. See John Burns, "British Muslim Leaders Propose 'Code of Conduct,'" *New York Times*, November 30, 2007, http://www.nytimes.com/2007/11/30/world/europe/30britain.html (accessed March 21, 2011).

3. Carl Zimmer, "In Games, an Insight into the Rules of Evolution," *New York Times*, July 31, 2007.

4. Richard Dawkins, *The Selfish Gene*, 2nd ed. (Oxford: Oxford University Press, 1989), p. 98.

5. Richard Dawkins, *The Greatest Show on Earth: The Evidence for Evolution* (New York: Free Press, 2009), p. 249.

6. Charles Darwin, *The Origin of Species*, 2nd ed. (New York: Gramercy, 1979), p. 93.

7. Darwin, *Descent of Man*, 2nd ed., p. 81.

8. Charles Darwin, *The Expression of Emotions in Man and Animals*, in *From So Simple a Beginning: The Four Great Books of Charles Darwin*, edited by Edward O. Wilson (New York: Norton, 2006), p. 1473.

9. Lisa. A. Parr and Bridget M. Waller, "Understanding Chimpanzee Facial Expression: Insights into the Evolution of Communication," *Social Cognitive and Affective Neuroscience* 1 (2006): 221–28.

10. Spencer Wells, *Pandora's Seed* (New York: Random House, 2010), p. 96.

11. Marc D. Hauser, *The Evolution of Communication* (Cambridge, MA: MIT Press, 1996), p. 265.

12. Jonathan H. Turner, *On the Origins of Human Emotion* (Stanford, CA: Stanford University Press, 2000), p. 20.

13. Ibid., p. 85. See also Edward C. Stewart, "Culture of the Mind," in *Culture in the Communication Age*, edited by James Lull (London: Routledge, 2001), pp. 9–30.

14. Karl von Frisch, *The Dance Language and Orientation of Bees* (Cambridge, MA: Belknap, 1967); Karl von Frisch, "Honeybees: Do They Use Direction and Distance Information Provided by Their Dancers?" *Science* 158 (1967): 1073–76.

15. John Maynard Smith and Eörs Szathmáry, *The Origins of Life* (Oxford: Oxford University Press, 1999), p. 133.

16. Hauser, *Evolution of Communication*, pp. 511–13.

17. R. M. Seyfarth and D. L. Cheney, "Signalers and Receivers in Animal Communication," *Annual Review of Psychology* 54 (2003): 168.

18. Christoph Grüter, M. Sol Balbuena, and Walter M. Farina, "Informational Conflicts Created by the Waggle Dance," *Proceedings of the Royal Society B: Biological Sciences* 275, no. 1640 (June 7, 2008), http://rspb.royalsocietypublishing.org/content/275/1640/1321.full (accessed April 1, 2009).

19. Edward O. Wilson, *Sociobiology: A New Synthesis* (Cambridge, MA: Harvard University Press, 1975), p. 180.

20. Darwin, *Expression of Emotion in Man and Animals*, p. 1314.

21. Steven R. Lindsay, *Handbook of Applied Dog Behavior and Training*, vol. 1 (Ames: Iowa State University Press, 2000), p. 377.

22. Darwin, *Expression of Emotion in Man and Animals*.

23. Michael Tomasello, *Origins of Human Communication* (Cambridge, MA: MIT Press, 2008), p. 55.

24. Nicholas Wade, *The Faith Instinct* (New York: Penguin, 2009), p. 71.

25. Tomasello, *Origins of Human Communication*, p. 344.

26. Bjorn Carey, "Chimps Prove Altruistic and Cooperative," Live Science, March 2, 2006, http://www.livescience.com/7076-chimps-prove-altruistic-cooperative.html (accessed February 12, 2010).

27. Without focusing on communication, Chris Stringer and Peter Andrews make this point about primate group life in *The Complete World of Human Evolution* (New York: Thames & Hudson, 2005).

28. Tomasello, *Origins of Human Communication*, p. 105.

29. Stringer and Andrews, *Complete World of Human Evolution*, p. 185.

30. Ibid., p. 189.

31. Herman Pontzer, David A. Raichlen, and Michael D. Sockol, "The Metabolic Cost of Walking in Humans, Chimpanzees, and Early Hominins," *Journal of Human Evolution* 56 (2009): 43–54.

32. Dennis M. Bramble and Daniel E. Lieberman. "Endurance Running and the Evolution of *Homo*," *Nature* 432 (2004): 345–52.

33. Stringer and Andrews, *Complete World of Human Evolution*, p. 208.

34. Louis Leakey, *Adam's Ancestors* (London: Methuen, 1934).

35. The examples given here were taken from two main sources: Benjamin B. Beck, *Animal Tool Behavior* (New York: Garland, STPM, 1980), and Donald R. Griffin, *Animal Thinking* (Cambridge, MA: Harvard University Press, 1984).

36. Julian K. Finn, Tom Tregenza, and Mark D. Norman, "Defensive Tool Use in a Coconut-Carrying Octopus," *Current Biology* 19 (2009): R1069–70.

37. Donald R. Griffin, *Animal Minds* (Chicago: University of Chicago Press, 2001).

38. Bob Holmes, "Did Prehistoric Chimps Use Stone Tools Too?" *New Scientist*, February 12, 2007, http://www.newscientist.com/article/dn11165-did-prehistoric-chimps-use-stone-tools-too.html (accessed April 16, 2007).

39. William C. McGrew, "Chimpanzee Technology," *Science* 328 (2010): 579–80.

40. Robert Jurmaine et al., *Introduction to Physical Anthropology* (Belmont, CA: Wadsworth, 2007).

41. John Tierney, "A World of Eloquence in an Upturned Palm," *New York Times*, August 28, 2007.

42. Jurmaine et al., *Introduction to Physical Anthropology*.

43. Shannon P. McPherron et al., "Evidence for Stone-Tool-Assisted Consumption of Animal Tissues before 3.39 Million Years Ago at Dikka, Ethiopia," *Nature* 466 (2010): 857–60.

44. Stringer and Andrews, *Complete World of Human Evolution*, p. 132.

45. Donald Johanson and Blake Edgar, *From Lucy to Language* (New York: Simon & Schuster, 2006), pp. 83, 106.

46. Nicholas Wade, "A Speech Gene Reveals Its Bossy Nature," *New York Times*, November 12, 2009.

47. Darwin, *Descent of Man*, 2nd ed., pp. 49, 633.

48. Clifford Nass and Li Gong, "Speech Interfaces from an Evolutionary Perspective: Social Psychological Research and Design Implications," *Association for Computing Machinery* 43 (2000): 36–43.

49. Stringer and Andrews, *Complete World of Human Evolution*, p. 226.

50. Ibid., p. 193.

51. Nicholas Wade, *Before the Dawn* (New York: Penguin, 2006), p. 81.

52. Some recent research calls into question the timing of the emigration from Africa, putting the time line closer to 125,000 years ago. See Simon J. Armitage et al., "The Southern Route 'Out of Africa': Evidence for an Early Expansion of Modern Humans into Arabia," *Science* 28 (2011): 453–56.

53. This summary draws from Johanson and Edgar, *From Lucy to Language*; Stringer and Andrews, *Complete World of Human Evolution*; and Wade, *Before the Dawn*.

54. Wade, *Before the Dawn*, p. 64.

55. Ibid., p. 8.

56. Michael S. Gazzaniga, *Nature's Mind* (New York: Basic Books, 1992), pp. 74–75.

57. Wade, *Before the Dawn*, p. 52.

58. Darwin, *Descent of Man*, 2nd ed., p. 90.

59. Noam Chomsky, *Aspects of the Theory of Syntax* (Cambridge, MA: MIT Press, 1965).

60. Steven Pinker, *The Language Instinct* (New York: William Morrow, 1994).

61. Michael Tomasello, "What Kind of Evidence Could Refute the UG Hypothesis?" *Studies in Language* 28 (2004): 642–44; Tomasello, *Origins of Human Communication*, p. 275.

62. Grant McCracken, *Culture and Consumption* (Bloomington: Indiana University Press, 1990), p. 63.

63. Mathias Osvath, "Spontaneous Planning for Future Stone Throwing by a Male Chimpanzee," *Current Biology* 19 (2009): R190–91.

64. Stringer and Andrews, *Complete World of Human Evolution*, p. 130.

65. Hauser, *Evolution of Communication*, p. 211.

66. Sherwood Washburn, *The Social Life of Early Man* (London: Methuen, 1962); Sherwood Washburn and V. Avis, "Evolution of Human Behavior," in *Behavior and Evolution*, edited by Anne Roe and George Simpson (New Haven, CT: Yale University Press, 1958), pp. 421–36.

67. Washburn, *Social Life of Early Man*, p. 153; Washburn and Avis, "Evolution of Human Behavior," p. 433.

68. Ibid.; see also Robert Ardrey, *African Genesis* (New York: Athenaeum, 1961).

69. A useful analysis of Louis Leakey's theory of human evolution can be found at http://www.talkorigins.org/faqs/homs/leakeydiag.html.

70. Raymond H. Dart, "The Predatory Transition from Ape to Man," *International Anthropological and Linguistic Review* 1 (1953): 201.

71. Matt Cartmill, *A View to a Death in the Morning: Hunting and Nature through History* (Cambridge, MA: Harvard University Press, 1993).

72. Ibid., p. 34.

73. C. K. Brain, *The Hunters or the Hunted? An Introduction to African Cave Taphonomy* (Chicago: University of Chicago Press, 1980); Donna Hart and Robert Sussman, *Man the Hunted* (Boulder, CO: Westview, 2005), p. 244; Lewis Binford, "Human Ancestors: Changing Views of Their Behavior," *Journal of Anthropological Archaeology* 4 (1985): 292–327; Lewis Binford, "Subsistence, A Key to the Past," in *Cambridge Encyclopedia of Human Evolution*, edited by Steve Jones, Robert Martin, and David Pilbeam (Cambridge: Cambridge University Press, 1992), pp. 365–68.

74. Stringer and Andrews, *Complete World of Human Evolution*, p. 190.

75. Hart and Sussman, *Man the Hunted.*

76. P. Jackson, "Man versus Man-Eater," in *Great Cats: Majestic Creatures of the Wild*, edited by John Seidensticker, Susan Lumpkin, and Francis Knight (Emmaus, PA: Rodale, 1991), pp. 212–13; Norimitsu Onishi, "Trying to Save Wild Tigers by Rehabilitating Them," *New York Times*, April 21, 2010, http://www.nytimes.com/2010/04/22/world/asia/22tigers.html (accessed December 23, 2010).

77. Hans Krukk, *The Spotted Hyena: A Study of Predation and Social Behavior* (Chicago: University of Chicago Press, 1972), p. 158.

78. Brain, *The Hunters or the Hunted?*; Hart and Sussman, *Man the Hunted.*

79. Wells, *Pandora's Seed*, pp. 111–13.

80. Sarah Blaffer Hrdy, *Mothers and Others: The Evolutionary Origins of Mutual Understanding* (Cambridge, MA: Harvard University Press, 2010).

81. David Sloan Wilson, *Evolution for Everyone* (New York: Delacorte, 2007), p. 154.

82. Darwin, *Descent of Man*, 2nd ed., p. 598.

83. Ibid., p. 513.

84. Donald E. Brown, *Human Universals* (Philadelphia: Temple University Press, 1991).

85. Claudia Dreifus, "Always Revealing, Human Skin Is an Anthropologist's Map," *New York Times*, January 9, 2007, http://www.nytimes.com/2007/01/09/science/09conv.html?sq=Always%20Revealing,%20Human%20Skin%20Is%20an%20Anthropologist%E2%80%99s%20Map:%20A%20Conversation%20with

%20Nina%20Jablonski&st=cse&adxnnl=1&scp=1&adxnnlx=1329154187-6g8VC9I5WmmYhv+paRQJBg (accessed June 29, 2009).

86. Ibid.

87. Christopher S. Henshilwood et al., "A 100,000-Year Old Ochre-Processing Workshop at Blombos Cave, South Africa," *Science* 14 (October 2011): 219–22.

88. Marian Vanhaeren et al., "Middle Paleolithic Shell Beads in Israel and Algeria," *Science* 312 (June 23, 2006): 1785–88.

89. Wells, *Pandora's Seed*, pp. 15–16.

90. Stringer and Andrews, *Complete World of Human Evolution*, p. 220; Johanson and Edgar, *From Lucy to Language*, p. 102.

91. Johanson and Edgar, *From Lucy to Language*, p. 99.

92. Nicholas J. Conard, "A Female Figurine from the Basal Aurignacion of Hohle Fels Cave in Southwestern Germany," *Nature* (May 14, 2009), http://www.urgeschichte.uni-tuebingen.de/fileadmin/downloads/Conard/Conard_Venus_Nature_2009.pdf (accessed May 30, 2009).

93. Nicholas J. Conard, Maria Malina, and Susanne C. Münzel, "New Flutes Document the Earliest Musical Tradition in Southwestern Germany," *Nature* (June 24, 2009), http://www.nature.com/nature/journal/v460/n7256/full/nature08169.html (accessed July 30, 2011).

94. Daniel Nettle and Helen Clegg, "Schizotypy, Creativity, and Mating Success in Humans," *Proceedings of the Royal Society B: Biological Sciences* 273 (2006): 611–15.

95. Stephen Jay Gould and Richard C. Lewontin, "The Spandrels of San Marco and the Panglossian Paradigm," *Proceedings of the Royal Society of London* 205 (1979): 281–88; Stephen Jay Gould, *The Structure of Evolutionary Theory* (Cambridge, MA: Harvard University Press, 2002).

96. Denis Dutton, *The Art Instinct* (New York: Bloomsbury, 2009), pp. 100, 136.

97. Steven Mithen, *The Singing Neanderthals* (Cambridge, MA: Harvard University Press, 2005).

98. Charles Darwin, *The Voyage of the Beagle*, in Wilson, *From So Simple a Beginning*, p. 219.

99. Ibid.

100. Darwin, *Origin of Species*, pp. 459–60.

101. Ibid.

CHAPTER 3. COMMUNICATING SEX

1. Facebook®, http://www.facebook.com/statistics (accessed September 1, 2011). Note: at time of publication, this page was inactive.

2. Zadie Smith, "The Social Network," *New York Review of Books* 25 (2010): 57–60.

3. Pew Internet and American Life Project, "On MySpace, Girls Seek Friends, Boys Flirt," January 7, 2007, http://pewinternet.org/Media-Mentions/2007/On-MySpace-girls-seek-friends-boys-flirt-study.aspx (accessed February 23, 2012).

4. Erasmus Darwin, *The Collected Writings of Erasmus Darwin: Zoonomia and the Laws of Organic Life* (Bristol, UK: Continuum, 2004).

5. Charles Lyell, *Principles of Geology* (London: John Murray, 1837).

6. Francis Darwin, ed., *Charles Darwin: His Life Told in an Autobiographical Chapter and in a Selected Series of Letters* (London: John Murray, 1902), p. 134.

7. Charles Darwin, *The Origin of Species*, 2nd ed. (New York: Gramercy, 1979).

8. Charles Darwin, with Nora Barlow, ed., *The Autobiography of Charles Darwin 1809–1882* (New York: Harcourt & Brace, 1959), p. 57.

9. Ibid., pp. 56–57.

10. Ibid., p. 57.

11. F. Darwin, *Charles Darwin: His Life Told in an Autobiographical Chapter*, p. 166.

12. John Maynard Smith, *The Theory of Evolution*, Canto ed., rev. (Cambridge: Cambridge University Press, 1993), p. 38.

13. Charles Darwin, *The Descent of Man*, 2nd ed. (Amherst, NY: Prometheus Books, 1998), p. 44, and Charles Darwin, *The Descent of Man, and Selection in Relation to Sex* (London: John Murray, 1871).

14. Olivia Judson, "An Evolve-By Date," Opinionator, *New York Times*, November 24, 2009, http://opinionator.blogs.nytimes.com/2009/11/24/an-evolve-by-date/ (accessed September 1, 2011).

15. Richard Dawkins, *The Blind Watchmaker: Why the Evidence of Evolution Reveals a Universe without Design* (New York: Norton, 1986).

16. Darwin, *Origin of Species*, p. 157.

17. Ibid., pp. 281, 444.

18. Ibid., p. 151.

19. The Lesser Rhea is now also known as Darwin's Rhea.

20. Dividing lines between species are never completely clear. For example, a species of frog found in Borneo in 2008 has no lungs and breathes through the skin. Some other amphibians have the same trait, but the discovery of a frog with this anatomical feature is unique. The frog's biological characteristics developed from an adaptation that was made to fast-moving streams in the area. David Bickford, Djoko Iskander, and Anggraini Barlian, "A Lungless Frog Discovered on Borneo," *Current Biology* 18 (2008): R374–75. Nature selects the total biology of each organism. No

single trait or combination of favorable traits ensures selection. See Jerry Fodor and Massimo Piatelli-Palmarini, *What Darwin Got Wrong* (New York: Farrar, Straus and Giroux, 2010).

21. Dawkins, *Blind Watchmaker*, p. 263.

22. John Maynard Smith and Eörs Szathmáry, *The Origins of Life* (Oxford: Oxford University Press, 1999), p. 2.

23. Richard Dawkins, *The Greatest Show on Earth: The Evidence for Evolution* (New York: Free Press, 2009).

24. Jerry Coyne, *Why Evolution Is True* (New York: Penguin, 2009), p. 119.

25. Richard Dawkins, "Afterword," presented to the London School of Economics and Political Science, March 16, 2006.

26. Richard Dawkins, "Darwin's Five Bridges," address given to Darwin Anniversary Festival, Cambridge, UK, July 6, 2009.

27. Nicholas Wade, "Still Evolving, Human Genes Tell New Story," *New York Times*, March 7, 2006, http://www.nytimes.com/2006/03/07/science/07evolve.html?scp=1&sq=Still%20Evolving,%20Human%20Genes%20Tell%20New%20Story&st=cse (accessed March 25, 2011).

28. Darwin, *Origin of Species*, p. 130.

29. Ibid., p. 175.

30. Ibid., p. 213.

31. Nicholas Wade, *Before the Dawn* (New York: Penguin, 2006), p. 174.

32. Darwin, *Descent of Man*, 2nd ed., p. 638.

33. Richard Dawkins, *The Selfish Gene*, 2nd ed. (Oxford: Oxford University Press, 1989), p. 43.

34. Smith and Szathmáry, *Origins of Life*, p. 87.

35. William D. Hamilton, *Narrow Roads of Gene Land: The Collected Papers of W. D. Hamilton, Volume 2: The Evolution of Sex* (Oxford: Oxford University Press, 2001).

36. Matthew R. Goddard, Charles Godfray, and Austin Burt, "Sex Increases the Efficacy of Natural Selection in Experimental Yeast Populations," *Nature* 434 (March 31, 2005): 636–40.

37. Donald Johanson and Blake Edgar, *From Lucy to Language* (New York: Simon & Schuster, 2006), p. 89.

38. Ibid.

39. Eric Alden Smith, "Why Do Good Hunters Have Higher Reproductive Success?" *Human Nature* 15 (2004): 343–64; Hillard Kaplan and Kim Hill, "Hunting Ability and Reproductive Success among Male Ache Foragers," *Current Anthropology* 26 (1985): 131–33; Craig B. Stanford, *The Hunting Apes: Meat Eating and the Origins of Human Behavior* (Princeton, NJ: Princeton University Press, 1999).

40. Ernst Mayr, *The Growth of Biological Thought: Diversity, Evolution, and Inheritance* (Cambridge, MA: Belknap, 1982).

41. Essentially the same food-for-sex strategy is employed by many males in modern cultures today in the guise of the dinner date. Rejecting the implicit terms of this custom has become something of a cause célèbre for some women.

42. Public Broadcasting Service, "Nature: What Females Want and Males Will Do," telecast nationally in the United States, April 6 and 13, 2008.

43. Darwin, *Descent of Man*, 2nd ed., p. 222.

44. Ibid., pp. 370, 362–63.

45. Public Broadcasting Service, "Nature: What Females Want and Males Will Do."

46. Jakob Bro-Jørgensen and Wiline M. Pangle, "Male Topi Antelopes Alarm Snort Deceptively to Retain Females for Mating," *American Naturalist* 176, no. 1 (July 2010), http://www.jstor.org/stable/10.1086/653078?&Search=yes&searchText=Deceptively&searchText=Mating&searchText=Snort&searchText=Females&searchText=Retain&searchText=Antelopes&searchText=Topi&searchText=Male&searchText=Alarm&list=hide&searchUri=%2Faction%2FdoBasicSearch%3FQuery%3DMale%2BTopi%2BAntelopes%2BAlarm%2BSnort%2BDeceptively%2Bto%2BRetain%2BFemales%2Bfor%2BMating%26acc%3Doff%26wc%3Don&prevSearch=&item=1&ttl=1&returnArticleService=showFullText (accessed October 11, 2010).

47. Jennifer Welsh, "8-Legged Sex Trick? Spiders Give Worthless Gifts, Play Dead," November 14, 2011, http://www.livescience.com/17010-spider-gifts-play-dead-mating.html (accessed February 5, 2012).

48. David P. Barash and Judith E. Lipton, *The Myth of Monogamy* (New York: Holt, 2002).

49. Malte E. Andersson, *Sexual Selection* (Princeton, NJ: Princeton University Press, 1994).

50. Dawkins, *Greatest Show on Earth*, p. 54.

51. John Krebs, "From Intellectual Plumbing to Arms Race," presented to the London School of Economics and Political Science, March 16, 2006.

52. Coyne, *Why Evolution Is True*, p. 158.

53. Ibid., p. 160.

54. Marc D. Hauser, *The Evolution of Communication* (Cambridge, MA: MIT Press, 1996), p. 29.

55. W. L. Thompson, "Agonistic Behavior in the House Finch, Part 1: Annual Cycle and Display Patterns," *Condor* 62 (1960): 245–71.

56. Denis Dutton, *The Art Instinct* (New York: Bloomsbury, 2009), p. 138.

57. Dawkins, *Greatest Show on Earth*, p. 136.

58. Eduardo Neiva, *Communication Games* (Berlin: Mouton de Gruyter, 2007).

59. John Maynard Smith and G. R. Price, "The Logic of Animal Conflict," *Nature* 246 (1973): 15–18.

60. Geoffrey Miller, *Spent: Sex, Evolution, and Consumer Behavior* (New York: Viking, 2009).

61. Bonnie Gabriel, *The Fine Art of Erotic Talk* (New York: Bantam Books, 1998).

62. Daniel Nettle and Helen Clegg, "Schizotypy, Creativity, and Mating Success in Humans," *Proceedings of the Royal Society B: Biological Sciences* 273, no. 1586 (2006): 611–15.

63. Susan Blackmore, *The Meme Machine* (Oxford: Oxford University Press, 1999), pp. 130–31.

64. Terence Kealey, "Why Do Men Find Big Lips and Little Noses So Sexy? I'll Paint You a Picture," *Times*, November 28, 2005, http://www.times.co.uk/tto/law/columnists/article/2048633ece (accessed August 12, 2009).

65. Darwin, *Descent of Man*, 2nd ed., p. 593.

66. Steven Mithen, *The Singing Neanderthals* (Cambridge, MA: Harvard University Press, 2005); Steven L. Kuhn and Mary C. Stiner, "What's a Mother to Do?" *Current Anthropology* 47, no. 6 (December 2006): 953–81.

67. Especially for men, to fail as a dancer indicates lack of sexual promise and threatens rejection. No matter what the age, ethnicity, or gender, people seek cultural comfort zones that reinforce their sexual potential and avoid situations like dance that call it into question. Many older people, for instance, enjoy getting on the dance floor when familiar hits from the fifties or sixties are playing but steadfastly refuse to move to the less familiar beat of rap or reggaeton.

68. Richard D. Alexander and Kayley M. Noonan, "Concealment of Ovulation: Parental Care and Human Social Evolution," in *Evolutionary Biology and Human Social Behavior*, edited by Napoleon A. Chagnon and William Irons (North Scituate, MA: Duxbury, 1979), pp. 436–53.

69. Other species that copulate frequently, such as bonobos, lions, goshawks, and white ibises, do so because of sperm completion resulting from females living in groups with several males. In the case of these animals, constant sex serves as a counterstrategy to increase the chances for reproduction and reduce the threat of raising another organism's offspring.

70. Christina M. Gomes and Christopher Boesch, "Wild Chimpanzees Exchange Meat for Sex in a Long-Term Basis," *PLoS ONE* 116 (2009): 1371.

71. Christopher Ryan and Cacilda Jethá, *Sex at Dawn: The Prehistoric Origins of Modern Sexuality* (New York: Harper, 2010).

72. Sewall Wright, "The Evolution of Dominance," *American Naturalist* 63 (1929): 556–61; Douglas J. Futuyma, *Evolutionary Biology*, 2nd ed. (Sunderland,

MA: Sinauer, 2009); Charlotte J. Avers, *Process and Pattern in Evolution* (Oxford: Oxford University Press, 1989).

73. Darwin, *The Origin of Species*, 6th ed. (London: John Murray, 1888), p. 412.

CHAPTER 4. COMMUNICATING CULTURE

1. Samuel P. Huntington, *The Clash of Civilizations and the Remaking of World Order* (New York: Simon & Schuster, 1996); Charles Darwin, *The Expression of Emotions in Man and Animals*, in *From So Simple a Beginning: The Four Great Books of Charles Darwin*, edited by Edward O. Wilson (New York: Norton, 2006), p. 1349.

2. Charles Darwin, *The Origin of Species*, 2nd ed. (New York: Gramercy, 1979), p. 113.

3. John Maynard Smith and Eörs Szathmáry, *The Origins of Life* (Oxford: Oxford University Press, 1999), p. 138.

4. John Maynard Smith, *Evolution and the Theory of Games* (Cambridge: Cambridge University Press, 1982); Richard Dawkins, *The Selfish Gene*, 2nd ed. (Oxford: Oxford University Press, 1976), p. 84.

5. Michael Tomasello, *The Origins of Human Communication* (Boston: MIT Press, 2008), pp. 213, 290.

6. Ibid., p. 209.

7. Richard Dawkins, *The Greatest Show on Earth: The Evidence for Evolution* (New York: Free Press, 2009), p. 402.

8. Tomasello, *The Origins of Human Communication*, p. 344.

9. Nicholas Wade, *Before the Dawn* (New York: Penguin, 2006), p. 34.

10. Donald Johanson and Blake Edgar, *From Lucy to Language* (New York: Simon & Schuster, 2006), p. 21.

11. Clifford Geertz, *The Interpretation of Cultures* (New York: Basic Books, 1973), p. 48.

12. Genes don't do this alone. Ribonucleic acid (RNA) molecules also influence the complex process of protein production.

13. Peter J. Richerson and Robert Boyd, *Not by Genes Alone* (Chicago: University of Chicago Press, 2005), p. 9.

14. Ibid.; the helpful term "internal cultural patterns" is from Thomas Sowell, *Race and Culture* (New York: Basic Books, 1994).

15. Steven Pinker and Paul Bloom, "Natural Language and Natural Selection," *Behavioral and Brain Science* 13 (1990): 707–84.

16. Jerry Coyne, *Why Evolution Is True* (New York: Penguin, 2009), p. 143.

17. John Hawks et al., "Recent Acceleration of Human Adaptive Evolution," *Proceedings of the National Academy of Sciences* 104 (2007): 20753–58.

18. Henry C. Harpending, *The 10,000 Year Explosion* (New York: Basic Books, 2009).

19. Spencer Wells, *Pandora's Seed* (New York: Random House, 2010).

20. Xin Yi, "Sequencing of 50 Human Exomes Reveals Adaptation to High Altitude," *Science* 329 (2010): 75–78.

21. Nabil Sabri Enattah et al., "Independent Introduction of Two Lactase Persistence Alleles into Human Populations Reflects Different Histories of Adaptation to Milk Culture," *American Journal of Human Genetics* 82 (2008): 57–72.

22. Dawkins, *Selfish Gene*, p. 20.

23. Ibid., pp. 189–201.

24. Ibid., p. 192.

25. Richard Dawkins, *The God Delusion* (New York: Houghton Mifflin, 2006), p. 191.

26. Richard Dawkins, "Afterword," address presented to the London School of Economics and Political Science, March 16, 2006.

27. Daniel Dennett, *Darwin's Dangerous Idea: Evolution and the Meaning of Life* (New York: Simon & Schuster, 1995).

28. Daniel Dennett, "Darwin and the Evolution of 'Why?'" address presented to the Darwin Anniversary Festival, Cambridge, UK, July 8, 2009.

29. Susan Blackmore, *The Meme Machine* (Oxford: Oxford University Press, 1999), p. 182.

30. Ibid., p. 17.

31. Paul Marsden, "Memetics and Social Contagion: Two Sides of the Same Coin?" *Journal of Memetics: Evolutionary Models of Information Transmission* 2 (1998), http://cfpm.org/jom-emit/1998/vol2/marsden_p.html (accessed November 30, 2010).

32. Blackmore, *The Meme Machine*, p. 4.

33. Kate Distin, *The Selfish Meme* (Cambridge: Cambridge University Press, 2005).

34. David L. Hull, *Science as a Process* (Chicago: University of Chicago Press, 1988); Robert Aunger, *The Electric Meme* (New York: Free Press, 2002).

35. Dawkins, *God Delusion*, pp. 196–200.

36. Darwin, *Origin of Species*, p. 201.

37. Dawkins, *God Delusion*, p. 198.

38. Ibid., p. 116.

39. Dennett, *Darwin's Dangerous Idea*, pp. 342–52.

40. Charles Darwin, *The Descent of Man*, 2nd ed. (Amherst, NY: Prometheus Books, 1998), p. 127.

41. The memetic theory of cultural transmission also fits with a more recent trend—the science of social contagion, which derives from the personal-influence research tradition of mid-nineteenth-century social psychology. Opinion leaders influence behavior through a multi-step flow of information and opinion filtering down from experts to followers on topics ranging from beauty tips and health aids to political candidates and consumer goods. The underlying premise of social contagion is evolutionarily sound: all humanity is interconnected, and our behavior inevitably affects the actions of others. The explanation that is offered about how these actions play out, however, is weak: good ideas, bad ideas, moods, and behaviors travel from person to person, sometimes indirectly, by an unspecified mechanism of influence. Individuals who make up a social network come to share emotional states like happiness, physical traits like being overweight, or personal habits like smoking. Person A might influence Person B directly so that both individuals come to share a trait. But Person A could also influence Person C, who is connected directly to Person B even if Person B is not affected and Persons A and C never meet. According to this view, the social network unleashes a mysterious power of its own. People converge in ways that exceed the power of any individual to affect another directly. See Nicholas A. Christakis and James H. Fowler, *Connected* (New York: Little, Brown, 2009).

42. Johanson and Edgar, *From Lucy to Language*, p. 83.

43. Alison Drain, "In Looking at the Cultured Ape, Researchers Learn Much about Humanity," American Association for the Advancement of Science, February 20, 2006, http://www.aaas.org/news/releases/2006/0220apes.shtml (accessed May 2, 2010).

44. Kristina Cawthon Lang, "Primate Fact Sheets: Gorilla Behavior," Primate Info Net, October 4, 2005, http://pin.primate.wisc.edu/factsheets/entry/gorilla (accessed November 5, 2011).

45. Steve Jones, *Darwin's Ghost: The Origin of Species Updated* (New York: Random House, 2000).

46. Wolfgang Wickler, *Mimicry in Plants and Animals* (New York: McGraw-Hill, 1968).

47. Richerson and Boyd, *Not by Genes Alone*, p. 116.

48. Ibid., p. 107.

49. Harold Garfinkel, *Studies in Ethnomethodology* (Englewood Cliffs, NJ: Prentice-Hall, 1967).

50. Judge for yourself by Googling her image.

51. Darwin, *Origin of Species*, p. 157.

52. Steven Johnson, *Where Good Ideas Come From* (New York: Riverhead Books, 2010), p. 166.

53. Parag Khanna, "Beyond City Limits," *Foreign Policy* (September 2010): 120–28.

54. Richerson and Boyd, *Not by Genes Alone*, p. 50.

55. Raymond Williams, *The Long Revolution* (New York: Columbia University Press, 1962).

56. Richard Lewontin, "It's Even Less in Your Genes," *New York Review of Books*, May 26, 2001, p. 23.

57. Lawrence Lessig, *Remix: Making Art and Commerce Thrive in the Hybrid Economy* (New York: Penguin, 2008).

58. See Toshie Takahashi, *Audience Studies: A Japanese Perspective* (New York: Routledge, 2010), for a summary of this argument.

59. James Lull, *Culture-on-Demand* (Oxford, UK: Blackwell, 2007); James Lull, ed., *Culture in the Communication Age* (London: Routledge, 2001).

60. Marshall McLuhan, *The Gutenberg Galaxy* (Toronto, ON: Toronto University Press, 1962); Marshall McLuhan, *Understanding Media* (New York: New American Library, 1964).

61. James Lull, *Media, Communication, Culture* (Cambridge, UK: Polity, 2000), p. 38.

62. George Dyson, *Darwin among the Machines* (New York: Basic Books, 1998); George Dyson, "Evolution of Technology," address to the National Aeronautics and Space Administration, Mountain View, California, October 19, 2009.

63. Johnson, *Where Good Ideas Come From*, p. 142.

64. David Wootton, *Galileo: Watcher of the Skies* (Princeton, NJ: Princeton University Press, 2010).

65. Darwin, *The Descent of Man* (Princeton, NJ: Princeton University Press, 1981) pp. 126–27.

66. Richard Dawkins, "The Genius of Charles Darwin," IWC Media, Channel 4, London, 2008; Richard Dawkins, "Randolph Nesse Interview," Richard Dawkins Foundation, March 16, 2009, http://richarddawkins.net/rdf_productions/randolph_nesse (accessed October 30, 2010).

67. Johanson and Edgar, *From Lucy to Language*, p. 111; Terrence W. Deacon, *The Symbolic Species* (London: Penguin, 1997).

68. Darwin, *Origin of Species*, p. 97.

69. United Nations Development Programme, *Human Development Report: Cultural Liberty in Today's Diverse World* (New York: Oxford University Press, 2004).

CHAPTER 5. COMMUNICATING MORALITY

1. Janet Browne, *Charles Darwin: A Biography, Vol. 2: The Power of Place* (New York: Knopf, 2002), p. 87.

2. Michael Shapiro, "Who Killed Pat Tillman?" AlterNet, June 13, 2007, http://www.alternet.org/world/53827 (accessed January 3, 2009).

3. Military authorities finally admitted to a cover-up of the circumstances surrounding Tillman's death. He had been killed by friendly fire. The investigating officer who made the crude comment about the Tillman family was later punished. (His remark that Pat Tillman would become worm dirt was technically correct. Becoming worm dirt will eventually be the officer's fate too, assuming he's buried. For his part, Charles Darwin was fascinated by worms and by the sinuous subterranean passages they create in worm dirt, just as he was enthralled by all the rest of nature's awesome diversity. His last book, *The Formation of Vegetable Mould through the Action of Worms* [London: John Murray, 1882], marked a return to this strong interest.) United States Army General Stanley McChrystal, who approved the investigating officer's report, later took command of American military forces in Afghanistan. His qualifications for making decisions in Afghanistan were questioned by some critics in light of his role in the Tillman affair.

4. Nicholas Wade, "Scientist Finds the Beginnings of Morality in Primate Behavior," *New York Times*, March 20, 2007.

5. Charles Darwin, *The Origin of Species*, 2nd ed. (New York: Gramercy, 1979), p. 453.

6. Marc D. Hauser, *Moral Minds* (New York: HarperCollins, 2007); Frans de Waal, *Our Inner Ape* (New York: Riverhead Books, 2005); Marco Iacoboni, *Mirroring People* (New York: Farrar, Straus and Giroux, 2008); Michael Tomasello, *Origins of Human Communication* (Cambridge, MA: MIT Press, 2008); Michael Tomasello, *Why We Cooperate* (Boston: MIT Press, 2009).

7. Richard Dawkins, *The God Delusion* (New York: Houghton Mifflin, 2006), p. 253.

8. George Schwab, *The Challenge of the Exception* (Berlin: Duncker & Humblot, 1989), p. 51.

9. Abu Zakariyya Yahya Ibn Sharaf an-Nawawi, *An-Nawawi's Forty Hadith: 2 CD Set* (Dar-us-Salam: Dar-us-Salam Publications, 2012).

10. Confucius, *The Analects*, translated by David Hinton (New York: Counterpoint, 1999).

11. Eduardo Neiva, *Communication Games* (Berlin: Mouton de Gruyter, 2007).

12. Charles Darwin, *The Descent of Man* (Princeton, NJ: Princeton University Press, 1981), p. 137.

13. Elliot Sober and David Sloan Wilson, *Unto Others: Evolution and the Psychology of Selfish Behavior* (Cambridge, MA: Harvard University Press, 1998); David Sloan Wilson, *Evolution for Everyone* (New York: Delacorte, 2007).

14. D. S. Wilson, *Evolution for Everyone*, p. 145.

15. Edward O. Wilson, *The Insect Societies* (Cambridge, MA: Harvard University Press, 1971); Edward O. Wilson, *Sociobiology: A New Synthesis* (Cambridge, MA: Harvard University Press, 1975); Bert Hölldobler and Edward O. Wilson, *The Superorganism: The Beauty, Elegance, and Strangeness of Insect Societies* (New York: Norton, 2008); V. C. Wynne-Edwards, *Animal Dispersion in Relation to Social Behavior* (Edinburgh, Scotland: Oliver and Boyd, 1962).

16. William D. Hamilton, *Narrow Roads of Gene Land: The Collected Papers of W. D. Hamilton, Volume I: The Evolution of Social Behavior* (Oxford, UK: W. H. Freeman/Spektrum, 1996), pp. 229–52.

17. Richard Dawkins, *The Greatest Show on Earth: The Evidence for Evolution* (New York: Free Press, 2009), p. 220.

18. George C. Williams, "Measurements of Consociation among Fish and Comments on the Evolution of Schooling," *Publications of the Museum*, Michigan State University, Biological Series 2/7 (1966): 149–84; George C. Williams, *Adaptation and Natural Selection: A Critique of Some Current Evolutionary Thought* (Princeton, NJ: Princeton University Press, 1996).

19. Tomasello, *Origins of Human Communication*, p. 181.

20. David Sloan Wilson, *Darwin's Cathedral* (Chicago: University of Chicago Press, 2002).

21. D. S. Wilson, *Evolution for Everyone*, p. 237.

22. G. R. Price, "Selection and Covariance," *Nature* 227 (1970): 520–21.

23. David Sloan Wilson and Edward O. Wilson, "Rethinking the Theoretical Foundation of Sociobiology," *Quarterly Review of Biology* 82 (2007): 327–48.

24. William D. Hamilton, "The Genetical Evolution of Social Behavior, Vols. 1 and 2," *Journal of Theoretical Biology* 7 (1964): 1–52.

25. John Maynard Smith and Eörs Szathmáry, *The Origins of Life* (Oxford: Oxford University Press, 1999), p. 129.

26. Darwin, *Origin of Species*, p. 258.

27. J. B. S. Haldane points out that we and our ancestors have no time to make such calculations: "It is clear that genes making for conduct of this kind would only have a chance of spreading in rather small populations where most of the children were fairly near relatives of the man who risked his life." J. B. S. Haldane, "Population Genetics," *New Biology* 18 (1955): 44.

28. Dustin R. Rubenstein and Irby J. Lovette, "Temporal Environmental Variability Drives the Evolution of Cooperative Breeding in Birds," *Current Biology* 17 (2007): 1414–19.

29. Martin A. Nowak, Corina E. Tarnita, and Edward O. Wilson, "The Evolution of Eusociality," *Nature* 466 (August 26, 2010): 1057–62.

30. Heini Hediger, "Proper Names in the Animal Kingdom," *Cellular and Molecular Sciences* 32 (1976): 1357–64. See also Thomas A. Sebeok, "Naming in Animals, with Reference to Playing," *Semiotic Inquiry* 1 (1981): 121–35. For a description of how sea lions recognize kin through the process of imprinting that utilizes hearing, vision, olfaction, and touch, see Ronald J. Schusterman, Colleen R. Kastak, and David Kastak, "The Cognitive Sea Lion: Meaning and Memory in the Laboratory and Nature," in *The Cognitive Animal: Empirical and Theoretical Perspectives in Animal Cognition*, edited by Marc Berkoff, Collin Allen, and Gordon M. Burghardt (Cambridge, MA: MIT Press, 2002), pp. 256–57.

31. Carol Kaesuk Yoon, "Loyal to Its Roots," *New York Times*, June 10, 2008.

32. Amotz Zahavi and Avishag Zahavi, *The Handicap Principle: A Missing Piece in Darwin's Puzzle* (Oxford: Oxford University Press, 1997).

33. Bryan D. Neff and Joanna S. Lister, "Genetic Life History Effects on Juvenile Survival in the Bluegill," *Journal of Evolutionary Biology* 20 (2007): 517–25.

34. Donald E. Brown, *Human Universals* (Philadelphia: Temple University Press, 1991).

35. Various estimates place mistaken paternity at around 10 percent in the United States, and at up to 30 percent in cases where paternity is contested. DNA testing for paternity has become big business. Males' desire to ensure that the children they raise are their own also leads to measures designed to control women in general.

36. Haldane, "Population Genetics," p. 44.

37. Robert L. Trivers, "The Evolution of Reciprocal Altruism," *Quarterly Journal of Biology* 46 (1971): 35–57.

38. Spencer Wells, *Pandora's Seed* (New York: Random House, 2010).

39. Zahavi and Zahavi, *Handicap Principle*, pp. 125–50.

40. Gerald S. Wilkinson, "Reciprocal Food Sharing in the Vampire Bat," *Nature* 389 (1984): 181–84.

41. Dawkins, *The God Delusion* (New York: Houghton Mifflin, 2006), p. 217.

42. British Broadcasting Corporation, "Galápagos: The Islands That Changed the World," 2007.

43. Charles Darwin, *The Structure and Distribution of Coral Reefs* (London: Smith, Elder, 1842).

44. Steven Johnson, *Where Good Ideas Come From* (New York: Riverhead Books, 2010), p. 245.

45. Carl Zimmer, "In Games, an Insight into the Rules of Evolution," *New York Times*, July 31, 2007.

46. Stephen G. Post, ed., *Altruism and Health* (Oxford: Oxford University Press, 2007).

47. Richard Wilkinson and Kate Pickett, *The Spirit Level: Why Great Equality Makes Societies Stronger* (New York: Bloomsbury, 2009).

48. Kimberly J. Hockings et al., "Chimpanzees Share Forbidden Fruit," *PLoS ONE*, September 9, 2007, http://www.plosone.org/article/info%3Adoi%2F10.1371%2Fjournal.pone.0000886 (accessed January 19, 2012).

49. Walter D. Koenig, Eric L. Walters, and Joey Haydock, "Variable Helper Effects, Ecological Conditions, and the Evolution of Cooperative Breeding in the Acorn Woodpecker," *American Naturalist* 178 (2011): 145–58.

50. Zahavi and Zahavi, *Handicap Principle*.

51. Francis Fukuyama, address given to the World Affairs Council, San Francisco, April 20, 2011.

52. Darwin, *Descent of Man*, p. 163.

53. Richard Dawkins, "The Genius of Charles Darwin," IWC Media, Channel 4, London, 2008.

54. Steven Mirsky, "What's Good for the Group," *Scientific American* (January 2009): 51.

55. Smith and Szathmáry, *Origins of Life*, p. 22.

56. Tomasello, *Origins of Human Communication*.

57. Peter J. Richerson and Robert Boyd, *Not by Genes Alone* (Chicago: University of Chicago Press, 2005); Iacoboni, *Mirroring People*.

58. Jeremy Rifkin, *The Empathic Civilization* (New York: Tarcher, 2009).

59. Marc Hauser, *Moral Minds* (New York: HarperCollins, 2007).

60. For a discussion of the main point made here, see Steven Pinker, "The Moral Instinct," *New York Times Magazine*, January 13, 2008, pp. 32–37, 52, 55–58. For a discussion of how Japanese social relations (*uchi*) are changing in the Communication Age, see Toshie Takahashi, *Audience Studies: A Japanese Perspective* (New York: Routledge, 2010).

61. Fernand Braudel, *Civilization and Capitalism: 15th–18th Centuries* (New York: Harper & Row, 1982–1984).

62. Fareed Zakaria, *The Future of Freedom* (New York: Norton, 2003), p. 45.

63. Stephen M. R. Covey, *The Speed of Trust* (New York: Free Press, 2006).

64. Richard Dawkins, *The Selfish Gene*, 2nd ed. (Oxford: Oxford University Press, 1989), p. 65.

65. Robert Trivers, *The Folly of Fools: The Logic of Deceit and Self-Deception in Human Life* (New York: Basic Books, 2012).

66. Darwin, *Descent of Man*, p. 633.

67. Ibid., p. 137.

CHAPTER 6. COMMUNICATING RELIGION

1. Charles Darwin, *The Descent of Man* (Princeton, NJ: Princeton University Press, 1981), p. 98. It's interesting that Darwin referred to his dog as "sensible," implying that humans aren't the only living creatures with the capacity for some degree of rational reflection.

2. Ibid.

3. David Quammen, *The Reluctant Mr. Darwin* (New York: Norton, 2006), p. 118.

4. Pew Global Attitudes Project, "World Publics Welcome Global Trade—But Not Immigration," Pew Research Center, October 4, 2007, http://www.pewglobal.org/2007/10/04/world-publics-welcome-global-trade-but-not-immigration/ (accessed April 18, 2009).

5. A. C. Grayling, "Atheists on Religion," address given to the London School of Economics, May 12, 2010.

6. David Carr and Tim Arango, "A Fox Chief at the Pinnacle of Media and Politics," *New York Times*, January 9, 2010, http://www.nytimes.com/2010/01/10/business/media/10ailes.html?pagewanted=all (accessed January 15, 2012).

7. Frank Newport, "Four in 10 Americans Believe in Strict Creationism," Gallup®, December 17, 2010, http://www.gallup.com/poll/145286/Four-Americans-Believe-Strict-Creationism.aspx (accessed July6, 2011).

8. T. A. Frail, "Poll: Americans Predict Life in 2050," *Smithsonian Magazine*, August 2010, http://www.smithsonianmag.com/specialsections/40th-anniversaryPoll-Americans-Predict-Life-in-2050.html (accessed June 27, 2011).

9. Jeffrey M. Jones, "Few Americans Oppose National Day of Prayer," Gallup®, May 5, 2010, http://www.gallup.com/poll/127721/Few-Americans-Oppose-National-Day-Prayer.aspx (accessed October 2, 2011).

10. Wendy Cadge and M. Daglian, "Blessings, Strength, and Guidance. Prayer Frames in a Hospital Prayer Book," *Poetics* 36 (2008): 358–73.

11. Pew Forum on Religion & Public Life, "The Religious Composition of the 112th Congress," Faith on the Hill, February 28, 2011, http://www.pewforum.org/Government/Faith-on-the-Hill--The-Religious-Composition-of-the-112th-Congress.aspx (accessed November 22, 2011).

12. Darwin, *Descent of Man*, p. 122.

13. Robert Wright, *The Evolution of God* (New York: Little, Brown, 2009), p. 433.

14. Immanuel Wallerstein, *The Decline of American Power* (New York: New Press, 2003), p. 102.

15. Ayaan Hirsi Ali, "Book Forum: Infidel," address presented to the American Enterprise Institute, Washington, DC, February 13, 2007.

16. Ibid.

17. Pew Global Attitudes Project, "Muslim Publics Divided on Hamas and Hezbollah," Pew Research Center, December 2, 2010, http://www.pewglobal.org/2010/12/02/muslims-around-the-world-divided-on-hamas-and-hezbollah/ (accessed August 5, 2011).

18. United Nations Development Programme, *Human Development Report: International Cooperation at a Crossroads* (New York: Oxford University Press, 2005); Sam Harris, *Letter to a Christian Nation* (New York: Knopf, 2006), pp. 43–44.

19. Ron Inglehart and Pippa Norris, "The True Clash of Civilizations," *Foreign Policy*, March–April 2003, p. 64.

20. Richard Dawkins, "Root of All Evil," IWC Media Documentary, Channel 4, London, 2006.

21. Darwin, *Descent of Man*, p. 126.

22. Charles Darwin, *The Expression of Emotions in Man and Animals*. In *From So Simple a Beginning: The Four Great Books of Charles Darwin*, edited by Edward O. Wilson (New York: Norton, 2006), p. 1468.

23. Donald E. Brown, *Human Universals* (Philadelphia: Temple University Press, 1991).

24. Darwin, *Descent of Man*, p. 126.

25. Richard Dawkins, *The God Delusion* (New York: Houghton Mifflin, 2006), pp. 189–90. Genetic drift is the tendency of genetic transmission to change randomly over time, leading to a gradual creation of patterns and characteristics that are not caused by the pressure of natural selection, especially in large populations where the effects of drift become more pronounced over time. Cultural drift refers to the same tendency, with respect to cultural features and behaviors.

26. Robert A. Hinde, *Religion and Darwinism* (London: British Humanist Association, 1997), p. 12.

27. Barbara J. King, *Evolving God* (New York: Doubleday, 2007).

28. Émile Durkheim, *The Elementary Forms of Religious Life* (New York: Oxford University Press, 2008).

29. David Sloan Wilson, *Evolution for Everyone* (New York: Delacorte, 2007), p. 259.

30. Ibid., p. 275.

31. Nicholas Wade, *The Faith Instinct* (New York: Penguin, 2009) p. 7.

32. Benedict Carey, "Do You Believe in Magic?" *New York Times*, January 23, 2007.

33. Sam Harris et al., "The Neural Correlates of Religious and Nonreligious Belief," *PLoS ONE*, June 29, 2009, http://www.plosone.org/article/info:doi/10.1371/journal.pone.0007272 (accessed February 9, 2011).

34. William James, *Varieties of Religious Experience: A Study of Human Nature* (Harmondsworth, UK: Penguin, 1982), p. 31.

35. Wade, *Faith Instinct*, p. 39.

36. This is the main argument made in ibid.

37. Ibid., p. 76.

38. Darwin, *Descent of Man*, p. 636.

39. Ibid., p. 97.

40. Hinde, *Religion and Darwinism*, p. 8.

41. Donald Johanson and Blake Edgar, *From Lucy to Language* (New York: Simon & Schuster, 2006), p. 102.

42. Chris Stringer and Peter Andrews, *The Complete World of Human Evolution* (New York: Thames & Hudson, 2005), pp. 218–19.

43. Wade, *Faith Instinct*, p. 94.

44. Ibid., p. 99.

45. Ibid., p. 88.

46. Darwin, *Expression of Emotions in Man and Animals*, p. 1388.

47. Christopher Hitchens, *God Is Not Great* (New York: Twelve, 2007), p. 227.

48. John Muir, quoted in a display at the Yosemite Museum, Yosemite National Park, California.

49. Thirty-eight percent of Americans feel this way: Newport, "Four in 10 Americans Believe in Strict Creationism.

50. Robert Wright, *The Evolution of God* (New York: Little, Brown, 2009).

51. Stephen Hawking and Leonard Mlodinow, *The Grand Design* (New York: Bantam Books, 2010), p. 41.

52. Wade, *Faith Instinct*, p. 154.

53. Ibid., p. 184.

54. Ludwig Schopp, *The Fathers of the Church*, Vol. 2, Internet Archive, http://www.archive.org/details/fathersofthechur009935mbp (accessed September 4, 2011).

55. The analogy was put forward by Russell in an article commissioned but never published by *Illustrated Magazine* in 1952. The original article can be accessed at Campaign for Philosophical Freedom: Bertrand Russell, "Is There a God?" http://www.cfpf.org.uk/articles/religion/br/br_god.html; Richard Dawkins, *A Devil's Chaplain* (New York: Mariner Books, 2004).

56. Alexandre Koyré, *From the Closed World to the Infinite Universe* (Radford, VA: Wilder, 2009).

57. Hawking and Mlodinow, *Grand Design*, p. 24.

58. Andrea C. Phelps et al., "Use of Religious Faith to Cope with Advanced Cancer Associated with Receiving Intensive Medical Care Near Death," *Journal of the American Medical Association* 301 (2009): 1140–47.

59. The image of Diego Maradona, the former coach of Argentina's national soccer team, furiously fingering the beads of his rosary as his team slipped toward elimination in the 2010 World Cup comes to mind, especially for Brazilians.

60. Robert Aunger, *The Electric Meme* (New York: Free Press, 2002).

61. Darwin, *Expression of the Emotions in Man and Animals*, p. 1443.

62. Ibid., p. 1459.

63. Ibid.

64. Dawkins, *God Delusion*, pp. 311–44.

65. Ibid., p. 206.

66. *Stanford Encyclopedia of Philosophy*, "Pascal's Wager," http://plato.stanford.edu/entries/pascal-wager/ (accessed January 2, 2010).

67. Dawkins, *God Delusion*, p. 105.

68. Ayaan Hirsi Ali, "From Islam to America: A Personal Journey through the Clash of Civilizations," World Affairs Council, San Francisco, May 26, 2010.

69. Hinde, *Religion and Darwinism*, pp. 11–12.

70. Ibid., p. 12.

71. Ibid., p. 11.

72. Ibid., p. 17.

73. A. C. Grayling, *Against All Gods* (London: Oberon, 2007), p. 28.

74. Vatican media said *Avatar* suggested that the worship of nature can replace religion. "Nature is no longer a creation to defend but a divinity to worship," according to Vatican Radio.

75. The nonpareil gift by God of his only begotten son to humankind functions for Christian authority the same way potlatch—a ceremonial feast among indigenous groups that live along the coast in North America's Great Northwest—serves the chief of the dominant tribe. At the end of the feast, the host bestows extremely valuable material gifts on guests from the other tribes. The recipients are forever rendered impotent by the weight of the overwhelming gifts. They surrender their souls to the giver. If they refuse to give in or can't return the favor in kind, the humiliation of receiving the gift will never leave them. See Aldona Jonaitis and Douglas Cole, *Chiefly Feasts: The Enduring Kwakiutl Potlatch* (Seattle: University of Washington Press, 1991). The reader may recognize something similar to potlatch in relationships with spouses, parents, or in-laws. The big, sparkling diamond wedding ring, for instance, can produce much the same psychological effect as a potlatch gift does on the recipient.

76. But even if the donor sincerely wants to remain anonymous, somebody (or bodies) always knows who the donor was and how much was given.

77. Or were they sweet white raisins? Translations of the Koran are not clear on this matter. And if the martyrs are female, then what?

78. Sigmund Freud, *The Future of an Illusion* (New York: Norton, 1961).

79. Gregory S. Paul, "The Big Religion Questions Finally Solved," *Free Inquiry*, (December–January 2008/2009): 34.

80. Steve Crabtree and Brett Pelham, "Religion Provides Emotional Boost to World's Poor," Gallup, March 6, 2009, http://www.gallup.com/poll/116449/Religion-Provides-Emotional-Boost-World-Poor.aspx (accessed March 16, 2011).

81. Darwin, *Descent of Man*, pp. 634–35.

82. Austin Dacey, *The Future of Blasphemy* (New York: Continuum, 2012).

83. Bertrand Russell, *Why I Am Not a Christian* (New York: Simon & Schuster, 1957), p. 47.

84. Dawkins, *God Delusion*, p. 306.

85. Terrence McNally, "Atheist Richard Dawkins on 'The God Delusion,'" AlterNet, January 18, 2007, http://www.alternet.org/story/46566/.

86. Eurobarometer, "Social Values, Science, and Technology," 2005, http://ec.europa.eu/public_opinion/archives/ebs/ebs_225_report_en.pdf (accessed January 3, 2010).

87. National Center for Social Research, "British Social Attitudes Survey, 25th Report," NatCen, January 2009, http://www.natcen.ac.uk/study/british-social-attitudes-25th-report (accessed September 10, 2011).

88. Steve Crabtree and Brett Pelham, "What Alabamians and Iranians Have in Common," Gallup, February 9, 2009, http://www.gallup.com/poll/114211/Alabamians-Iranians-Common.aspx (accessed September 10, 2011).

89. Frank Newport, "Near Record High See Religion Losing Its Influence in America," Gallup, December 29, 2010, http://www.gallup.com/poll/145409/Near-Record-High-Religion-Losing-Influence-America.aspx (accessed June 1, 2011).

90. Pew Forum on Religion & Public Life, "U.S. Religious Landscape Survey," June 23, 2008, http://religions.pewforum.org/reports (accessed December 1, 2008); Pew Forum on Religion & Public Life, "Faith in Flux: Changes in Religious Affiliation in the U.S.," April 27, 2009, http://www.pewforum.org/Faith-in-Flux-Changes-in-Religious-Affiliation-in-the-US.aspx (accessed October 30, 2009).

91. Newport, "Four in 10 Americans Believe in Strict Creationism."

92. Pew Forum on Religion & Public Life, "Many Americans Mix Multiple Faiths," December 9, 2009, http://www.pewforum.org/Other-Beliefs-and-Practices/Many-Americans-Mix-Multiple-Faiths.aspx (accessed January 30, 2012).

93. Juliana Menasce Horowitz, "Declining Support for bin Laden and Suicide Bombing," Pew Research Center Publications, September 10, 2009, http://pewresearch.org/pubs/1338/declining-muslim-support-for-bin-laden-suicide-bombing (accessed August 1, 2011); Pew Global Attitudes Project, "Mixed Views of Hamas and Hezbollah in Largely Muslim Nations," Pew Research Center, February 4, 2010, http://www.pewglobal.org/2010/02/04/mixed-views-of-hamas-and-hezbollah-in-largely

-muslim-nations/ (accessed June 18, 2011); Pew Global Attitudes Project, December 2, 2010, "Muslim Publics Divided on Hamas and Hezbollah," http://www.pewglobal.org/2010/12/02/muslims-around-the-world-divided-on-hamas-and-hezbollah/ (accessed October 18, 2011).

94. Pew Global Attitudes Project, "Mixed Views of Hamas and Hezbollah."

95. Ayaan Hirsi Ali, *Infidel* (New York: Free Press, 2007), p. 64.

96. Robert F. Worth, "Arab TV Tests Societies' Limits with Depictions of Wine, Sex, and Equality," *New York Times*, September 27, 2008.

97. George Basalla, *The Evolution of Technology* (Cambridge: Cambridge University Press, 1988), p. 131.

98. Peter J. Richerson and Robert Boyd, *Not by Genes Alone* (Chicago: Chicago University Press, 2005), pp. 209–10; Thomas Sowell, *Race and Culture* (New York: Basic Books, 1994), pp. 32–61.

99. Robert D. Putnam and David E. Campbell, *American Grace: How Religion Divides and Unites Us* (New York: Simon & Schuster, 2010).

100. Hawking and Mlodinow, *The Grand Design*, p. 17.

101. The symbolic power of this kind of religious mythology and iconography is astounding. When a murderous rampage left eight people dead in Akron, Ohio, in 2011, more than one hundred people posted condolences to the family on Facebook®, including one woman who saw a double rainbow after the shooting: "Your double rainbow sent from heaven did not go unnoticed. It was truly a sign from God that those taken are OK."

102. Stated in the gallery of the British Museum exhibit "Hajj: Journey to the Heart of Islam," January 26–April 15, 2012, London.

103. Hawking and Mlodinow, *The Grand Design*; Brian Greene, *The Hidden Reality* (New York: Alfred A. Knopf, 2011), pp. 8–9, 153, 165.

CHAPTER 7. COMMUNICATING CHANGE

1. Peter Sloterdijk, *Le Palais de Cristal* (Paris: Maren Sell Éditeurs, 2006).

2. In 1750, Great Britain granted ten patents for new inventions. The number of patents grew so steadily that by 1851, some 455 patents were given annually. See H. I. Dutton, *The Patent System and Inventive Activity during the Industrial Revolution 1750–1852* (Manchester, UK: Manchester University Press, 1984), p. 2.

3. George Basalla, *The Evolution of Technology* (Cambridge: Cambridge University Press, 1988), p. 77.

4. William Easterly, "Reinventing the Wheel: Why No-Tech Ancient Civilizations Still Can't Catch Up," *Foreign Policy* (November 2010): 44–45.

5. Kevin Kelly, *What Technology Wants* (New York: Viking, 2010), pp. 11, 15, 38.

6. Steven Johnson, *Where Good Ideas Come From* (New York: Riverhead Books, 2010), p. 50.

7. Paul Kennedy, *Preparing for the Twenty-First Century* (New York: Random House, 1993), p. 42.

8. Thomas R. Malthus, *An Essay on Population, or A View of Its Past and Present Effects on Human Happiness* (London: John Murray, 1798).

9. N. F. R. Crafts, *The Economic Growth during the Industrial Revolution* (Oxford, UK: Clarendon, 1985), p. 144.

10. T. S. Ashton, *The Industrial Revolution: 1760–1830* (London: Oxford University Press, 1962), p. 25.

11. Basalla, *Evolution of Technology*, p. 110.

12. Ibid., pp. 207–208. Marx, of course, insisted that we aren't just what we make; we are also defined by how we produce goods. The Industrial Revolution was still gathering momentum when Marx and Friedrich Engels wrote *The Communist Manifesto*. It is part of Marx's genius that what he glimpsed was not yet a tidal wave. Looking at *The Communist Manifesto*—a masterpiece of political agitation, theoretical concision, and persuasive bravado—one can see that Marx posited technological transformations, including communications, as the force that would remake economic life and change society. Karl Marx and Friedrich Engels, *The Communist Manifesto* (New York: Monthly Review, 1964), original citation: Karl Marx and Friedrich Engels, *Manifest der Kommunistischen Partei* (London: Communist League, 1848).

13. Marx and Engels, *Communist Manifesto*, p. 39.

14. Ibid.

15. Adam Smith, *The Wealth of Nations* (New York: Bantam, 2003).

16. Janet Browne, *Darwin's Origin of Species: A Biography* (New York: Atlantic Books, 2006), p. 54.

17. Scott Page, *The Difference: How the Power of Diversity Creates Better Groups, Firms, Schools, and Societies* (Princeton, NJ: Princeton University Press, 2007); G. Pascal Zachary, *The Global Me* (New York: Public Affairs, 2000).

18. Joseph A. Schumpeter, *Capitalism, Socialism, and Democracy* (New York: Routledge, 2006).

19. Basalla, *Evolution of Technology*, p. 208.

20. Charles Darwin, *The Origin of Species*, 2nd ed. (New York: Gramercy, 1979), p. 393.

21. James Lull, *Culture-on-Demand* (Oxford, UK: Wiley-Blackwell, 2007), pp. 195–96.

22. Jerry A. Coyne, *Why Evolution Is True* (New York: Penguin, 2009), pp. 121–22.

23. Nicholas Wade, *Before the Dawn* (New York: Penguin, 2006), p. 267.

24. Steven A. LeBlanc, *Constant Battles* (New York: St. Martin's, 2003), p. 8.

25. Human Security Centre, *Human Security Report 2005: War and Peace in the 21st Century* (New York: Oxford University Press, 2006). See also Steven Pinker, *The Better Angels of our Nature: Why Violence Has Declined* (New York: Viking, 2011).

26. Paul Seabright, *The Company of Strangers* (Princeton, NJ: Princeton University Press, 2004), p. 28.

27. Larry J. Diamond, *The Spirit of Democracy* (New York: Holt, 2008).

28. William J. Bernstein, *The Birth of Plenty* (New York: McGraw-Hill, 2010).

29. Vali Nasr, *Forces of Fortune: Rise of the New Muslim Middle Class and What It Will Mean for Our World* (New York: Free Press, 2009).

30. Frans de Waal, *Our Inner Ape* (New York: Riverhead Books, 2005).

31. Steven Pinker, "A History of (Non) Violence," *Foreign Policy* (December 2011), p. 86.

32. Charles Darwin, *The Descent of Man* (Princeton, NJ: Princeton University Press, 1981), p. 120.

33. Richard Dawkins, *The God Delusion* (New York: Houghton Mifflin, 2006), p. 221.

34. Peter J. Richerson and Robert Boyd, *Not by Genes Alone* (Chicago: University of Chicago Press, 2005), p. 230.

35. Sally McBrearty and Alison S. Brooks, "The Revolution That Wasn't: A New Interpretation of the Origin of Modern Human Behavior," *Journal of Human Evolution* 39 (2000): 453–563.

36. Wade, *Before the Dawn*, p. 234.

37. Ibid., pp. 140, 180.

38. Lynn Hunt, *Inventing Human Rights* (New York: Norton, 2007).

39. Michael Ignatieff, "Is Nothing Sacred? The Ethics of Television," *Daedalus* 114 (1985): 58.

40. Darwin, *Descent of Man*, p. 101.

41. Ibid., pp. 126–27.

42. Ibid., p. 101.

43. Salman Rushdie, comment on "Bill Moyers on Faith and Reason," Public Broadcasting Service, June 23, 2006.

44. Kwame Appiah, *Cosmopolitanism* (New York: Norton, 2006).

45. Robert Putnam, "*E Pluribus Unum*: Diversity and Community in the 21st Century," *Scandinavian Political Studies* 30 (2007): 137–74.

46. John Maynard Smith and Eörs Szathmáry, *The Origins of Life* (Oxford: Oxford University Press, 1999).

47. John Kenneth Galbraith, *The New Industrial State* (Harmondsworth, UK: Penguin/Hamish Hamilton, 1967).

48. Darwin, *Origin of Species*, p. 237.

49. Thomas C. Schelling, *The Strategy of Conflict* (Cambridge, MA: Harvard University Press, 1960).

50. Lull, *Culture-on-Demand*, p. 134.

51. James Lull and Stephen Hinerman, eds., *Media Scandals* (Cambridge: Polity, 1997).

52. Lull, *Culture-on-Demand*, pp. 133–50.

53. David Berlo, *The Process of Communication* (New York: Holt, Rinehart and Winston, 1960).

54. Manuel Castells, *Communication Power* (Oxford: Oxford University Press, 2009), p. 416.

55. Joseph S. Nye Jr., *The Future of Power* (New York: PublicAffairs Books, 2011).

56. See, for example, James Lull, *China Turned On: Television, Reform, and Resistance* (London: Routledge, 1991).

57. Bernard Lewis, *What Went Wrong?* (London: Weidenfeld & Nicolson, 2002).

58. James Lull, ed., *World Families Watch Television* (Newbury Park, CA: Sage, 1988); Robert Jensen and Emily Oster, "The Power of TV: Cable Television and Women's Status in India," *Journal of Economics* 124 (2009): 1057–94.

59. "Viral Video: A YouTube Is Born for the Arab World," *Newsweek*, January 22, 2007, p. 8.

60. Pew Global Attitudes Project, "Global Digital Communication: Texting, Social Networking Popular Worldwide," Pew Research Center, December 20, 2011, http://www.pewglobal.org/2011/12/20/global-digital-communication-texting-social-networking-popular-worldwide/ (accessed January 8, 2012).

61. Pew Global Attitudes Project, "Global Publics Embrace Social Networking," Pew Research Center, December 15, 2010, http://www.pewglobal.org/2010/12/15/global-publics-embrace-social-networking/ (accessed March 25, 2011).

62. As noted earlier, in key respects communications advances also widen the gaps between social groups. This is a serious issue but should not detract from the point being made here.

63. Niall Ferguson, "The Mash of Civilizations," Daily Beast, April 10, 2011, http://www.thedailybeast.com/newsweek/2011/04/10/the-mash-of-civilizations.html (accessed August 25, 2011).

64. Spencer Wells, *Pandora's Seed* (New York: Random House, 2010), p. xv.

65. Darwin, *Descent of Man*, p. 233.

66. Charles Darwin, *The Expression of Emotions in Man and Animals*, in *From So Simple a Beginning*, edited by Edward O. Wilson (New York: Norton, 2006), p. 1473.

67. Darwin, *Origin of Species,* p. 215.

68. Pew Global Attitudes Project, "The Global Middle Class," Pew Research Center, February 12, 2009, http://www.pewglobal.org/2009/02/12/the-global-middle-class/ (accessed December 23, 2010).

69. Richard Dawkins, *The Selfish Gene*, 2nd ed. (Oxford: Oxford University Press, 1989), pp. 200–201.

70. Charles Darwin, *The Voyage of the Beagle*, in Wilson, *From So Simple a Beginning* (New York: Norton, 2006), pp. 426–27.

BIBLIOGRAPHY

Alexander, Richard D., and Kayley M. Noonan. "Concealment of Ovulation: Parental Care and Human Social Evolution." In *Evolutionary Biology and Human Social Behavior*. Edited by Napoleon A. Chagnon and William Irons, pp. 436–53. North Scituate, MA: Duxbury, 1979.

Andersson, Malte E. *Sexual Selection*. Princeton, NJ: Princeton University Press, 1994.

an-Nawawi, Abu Zakariyya Yahya Ibn. *An-Nawawi's Forty Hadith: 2 CD Set*. Dar-us-Salam: Dar-us-Salam Publications, 2012.

Appiah, Kwame. *Cosmopolitanism*. New York: Norton, 2006.

Ardrey, Robert. *African Genesis*. New York: Athenaeum, 1961.

Armitage, Simon J., Sabah A. Jasim, Anthony E. Marks, Adrian G. Parker, Vitaly I. Usik, and Hans-Peter Uerpmann. "The Southern Route 'Out of Africa': Evidence for an Early Expansion of Modern Humans into Arabia." *Science* 28 (2011): 453–56.

Ashton, T. S. *The Industrial Revolution: 1760–1830*. London: Oxford University Press, 1962.

Associated Press. "Kansas Board Boosts Evolution Education," MSNBC, February 14, 2007. http://www.msnbc.msn.com/id/17132925/ns/technology_and_science/t/kansas-board-boosts-evolution-education/#.Ty_zucU9lm0 (accessed March 15, 2007).

Aunger, Robert. *The Electric Meme*. New York: Free Press, 2002.

Avers, Charlotte J. *Process and Pattern in Evolution*. Oxford: Oxford University Press, 1989.

Barash, David P., and Judith E. Lipton. *The Myth of Monogamy*. New York: Holt, 2002.

Basalla, George. *The Evolution of Technology*. Cambridge: Cambridge University Press, 1988.

Bawer, Bruce. *Surrender*. New York, Anchor, 2010.

———. *While Europe Slept*. New York: Anchor, 2007.

Beck, Benjamin B. *Animal Tool Behavior*. New York: Garland, STPM, 1980.

Berlo, David. *The Process of Communication*. New York: Holt, Rinehart and Winston, 1960.

Berman, Paul. *The Flight of the Intellectuals*. New York: Melville House, 2010.

Bernstein, William J. *The Birth of Plenty*. New York: McGraw-Hill, 2010.

Bickford, David, Djoko Iskander, and Anggraini Barlian. "A Lungless Frog Discovered on Borneo." *Current Biology* 18 (2008): R374–75.

Binford, Lewis. "Human Ancestors: Changing Views of Their Behavior." *Journal of Anthropological Archaeology* 4 (1985): 292–327.

———. "Subsistence: A Key to the Past." In *Cambridge Encyclopedia of Human Evolution*. Edited by Steve Jones, Robert Martin, and David Pilbeam, pp. 365–68. Cambridge: Cambridge University Press, 1992.

Blackmore, Susan. *The Meme Machine*. Oxford: Oxford University Press, 1999.

Bradt, Steve. "Molecular Analysis Confirms T. Rex's Evolutionary Link to Birds." *Harvard Gazette*, April 4, 2008. http://news.harvard.edu/gazette/story/2008/04/molecular-analysis-confirms-t-rexs-evolutionary-link-to-birds/ (accessed May 14, 2010).

Brain, C. K. *The Hunters or the Hunted? An Introduction to African Cave Taphonomy*. Chicago: University of Chicago Press, 1980.

Bramble, Dennis M., and Daniel E. Lieberman. "Endurance Running and the Evolution of *Homo*." *Nature* 432 (2004): 345–52.

Braudel, Fernand. *Civilization and Capitalism: 15th–18th Centuries*. New York: Harper & Row, 1982–1984.

British Broadcasting Corporation. "Galápagos: The Islands That Changed the World." 2007.

Bro-Jørgensen, Jakob, and Wiline M. Pangle. "Male Topi Antelopes Alarm Snort Deceptively to Retain Females for Mating." *American Naturalist* 176, no. 1 (July 2010). http://www.jstor.org/stable/10.1086/653078?&Search=yes&searchText=Deceptively&searchText=Mating&searchText=Snort&searchText=Females&searchText=Retain&searchText=Antelopes&searchText=Topi&searchText=Male&searchText=Alarm&list=hide&searchUri=%2Faction%2FdoBasicSearch%3FQuery%3DMale%2BTopi%2BAntelopes%2BAlarm%2BSnort%2BDeceptively%2Bto%2BRetain%2BFemales%2Bfor%2BMating%26acc%3Doff%26wc%3Don&prevSearch=&item=1&ttl=1&returnArticleService=showFullText (accessed October 11, 2010).

Brown, Donald E. *Human Universals*. Philadelphia: Temple University Press, 1991.

Browne, Janet. *Charles Darwin: A Biography, Volume 1: Voyaging*. New York: Knopf, 1995.

———. *Charles Darwin: A Biography, Volume 2: The Power of Place*. New York: Knopf, 2002.

———. *Darwin's Origin of Species: A Biography*. New York: Atlantic Books, 2006.

Burns, John. "British Muslim Leaders Propose 'Code of Conduct.'" *New York*

Times, November 30, 2007. http://www.nytimes.com/2007/11/30/world/europe/30britain.html (accessed March 21, 2011).

Cadge, Wendy, and M. Daglian. "Blessings, Strength, and Guidance: Prayer Frames in a Hospital Prayer Book." *Poetics* 36 (2008): 358–73.

Caldwell, Christopher. *Reflections on the Revolution in Europe*. New York: Anchor, 2010.

Carey, Benedict. "Do You Believe in Magic?" *New York Times*, January 23, 2007.

Carey, Bjorn. "Chimps Prove Altruistic and Cooperative." Live Science, March 2, 2006. http://www.livescience.com/7076-chimps-prove-altruistic-cooperative (accessed February 12, 2010).

Carr, David, and Tim Arango. "A Fox Chief at the Pinnacle of Media and Politics." *New York Times*, January 9, 2010. http://www.nytimes.com/2010/01/10/business/media/10ailes.html?pagewanted=all (accessed January 15, 2012).

Cartmill, Matt. *A View to Death in the Morning: Hunting and Nature through History*. Cambridge, MA: Harvard University Press, 1993.

Castells, Manuel. *Communication Power*. Oxford: Oxford University Press, 2009.

Cawthon Lang, Kristina. "Primate Fact Sheets: Gorilla Behavior." Primate Info Net, October 4, 2005. http://pin.primate.wisc.edu/factsheets/entry/gorilla (accessed November 5, 2011).

Chiappe, Luis M. *Glorified Dinosaurs*. New York: Wiley-Liss, 2007.

Chomsky, Noam. *Aspects of the Theory of Syntax*. Cambridge, MA: MIT Press, 1965.

Christakis, Nicholas A., and James H. Fowler. *Connected*. New York: Little, Brown, 2009.

Conard, Nicholas J. "A Female Figurine from the Basal Aurignacion of Hohle Fels Cave in Southwestern Germany." *Nature* (May 14, 2009). http://www.urgeschichte.uni-tuebingen.de/fileadmin/downloads/Conard/Conard_Venus_Nature_2009.pdf (accessed May 30, 2009).

Conard, Nicholas J., Maria Malina, and Susanne C. Münzel. "New Flutes Document the Earliest Musical Tradition in Southwestern Germany." *Nature* (June 24, 2009). http://www.nature.com/nature/journal/v460/n7256/full/nature08169 (accessed July 30, 2011).

Confucius. *The Analects*. Translated by David Hinton. New York: Counterpoint, 1999.

Covey, Steven M. R. *The Speed of Trust*. New York: Free Press, 2006.

Coyne, Jerry A. *Why Evolution Is True*. New York: Penguin, 2009.

Crabtree, Steve, and Brett Pelham. "Religion Provides Emotional Boost to World's Poor." Gallup, March 6, 2009. http://www.gallup.com/poll/116449/Religion-Provides-Emotional-Boost-World-Poor.aspx *(accessed* March 16, 2011).

———. "What Alabamians and Iranians Have in Common." Gallup, February 9,

2009. http://www.gallup.com/poll/114211/Alabamians-Iranians-Common.aspx (accessed September 10, 2011).

Crafts, N. F. R. *The Economic Growth during the Industrial Revolution*. Oxford, UK: Clarendon, 1985.

Dacey, Austin. *The Future of Blasphemy*. New York: Continuum, 2012.

Dart, Raymond H. "The Predatory Transition from Ape to Man." *International Anthropological and Linguistic Review* 1 (1953): 201–18.

Darwin, Charles. *The Descent of Man*. Princeton, NJ: Princeton University Press, 1981.

———. *The Descent of Man*. 2nd ed. Amherst, NY: Prometheus Books, 1998.

———. *The Expression of Emotions in Man and Animals*. In *From So Simple a Beginning*. Edited by Edward O. Wilson. New York: Norton, 2006.

———. *The Formation of Vegetable Mould through the Action of Worms*. London: John Murray, 1882.

———. *The Origin of Species*. 2nd ed. New York: Gramercy, 1979.

———. *The Origin of Species*. 6th ed. London: John Murray, 1888.

———. *The Structure and Distribution of Coral Reefs*. London: Smith, Elder, 1842.

———. *The Voyage of the Beagle*. In *From So Simple a Beginning*. Edited by Edward O. Wilson. New York: Norton, 2006.

Darwin, Charles, with Nora Barlow, ed. *The Autobiography of Charles Darwin: 1809–1882*. New York: Harcourt & Brace, 1959.

Darwin, Erasmus. *The Collected Writings of Erasmus Darwin: Zoonomia and the Laws of Organic Life*. Bristol, UK: Continuum, 2004.

Darwin, Francis, ed. *Charles Darwin: His Life Told in an Autobiographical Chapter and in a Selected Series of Letters*. London: John Murray, 1902.

Dawkins, Richard. "Afterword." Presented to the London School of Economics and Political Science, March 16, 2006.

———. *The Blind Watchmaker: Why the Evidence of Evolution Reveals a Universe without Design*. New York: Norton, 1986.

———. "Darwin's Five Bridges." Address given to Darwin Anniversary Festival, Cambridge, UK, July 6, 2009.

———. *A Devil's Chaplain*. New York: Mariner Books, 2004.

———. "The Genius of Charles Darwin." IWC Media, Channel 4, London, 2008.

———. *The God Delusion*. New York: Houghton Mifflin, 2006.

———. *The Greatest Show on Earth: The Evidence for Evolution*. New York: Free Press, 2009.

———. *The Magic of Reality: How We Know What's Really True*. New York: Free Press, 2011.

———. "Randolph Nesse Interview." Richard Dawkins Foundation, March 16,

2009. http://richarddawkins.net/rdf_productions/randolph_nesse (accessed October 30, 2010).

———. "Root of All Evil?" IWC Media documentary, Channel 4, London, 2006.

———. *The Selfish Gene*. 2nd ed. Oxford: Oxford University Press, 1989.

Deacon, Terrence W. *The Symbolic Species*: London: Penguin, 1997.

Dennett, Daniel. "Darwin and the Evolution of 'Why?'" Address presented to the Darwin Anniversary Festival, Cambridge, UK, July 8, 2009.

———. *Darwin's Dangerous Idea: Evolution and the Meaning of Life*. New York: Simon & Schuster, 1995.

Desmond, Adrian, and James Moore. *Darwin's Sacred Cause*. New York: Houghton Mifflin Harcourt, 2009.

de Waal, Frans. *Our Inner Ape*. New York: Riverhead Books, 2005.

Diamond, Larry J. *The Spirit of Democracy*. New York: Holt, 2008.

Distin, Kate. *The Selfish Meme*. Cambridge: Cambridge University Press, 2005.

Drain, Alison. "In Looking at the Cultured Ape, Researchers Learn Much about Humanity." American Association for the Advancement of Science, February 20, 2006. http://www.aaas.org/news/releases/2006/0220apes.shtml (accessed May 2, 2010).

Dreifus, Claudia. "Always Revealing, Human Skin Is an Anthropologist's Map." *New York Times*, January 9, 2007. http://www.nytimes.com/2007/01/09/science/09conv.html?sq=Always%20Revealing,%20Human%20Skin%20Is %20an%20Anthropologist%E2%80%99s%20Map:%20A%20Conversation %20with%20Nina%20Jablonski&st=cse&adxnnl=1&scp=1&adxnnlx=1329154187-6g8VC9I5WmmYhv+paRQJBg (accessed June 29, 2009).

Durkheim, Émile. *The Elementary Forms of Religious Life*. New York: Oxford University Press, 2008.

Dutton, Denis. *The Art Instinct*. New York: Bloomsbury, 2009.

Dutton, H. I. *The Patent System and Inventive Activity during the Industrial Revolution: 1750–1852*. Manchester, UK: Manchester University Press, 1984.

Dyson, George B. *Darwin among the Machines*. New York: Basic Books, 1998.

———. "Evolution of Technology." Address to the National Aeronautics and Space Administration, Mountain View, CA, October 19, 2009.

Easterly, William. "Reinventing the Wheel: Why No-Tech Ancient Civilizations Still Can't Catch Up." *Foreign Policy* (November 2010): 44–45.

Edgerton, David. *The Shock of the Old: Technology and Global History since 1900*. Oxford: Oxford University Press, 2007.

Enattah, Nabril Sabri, Tine K. G. Jensen, Mette Nielsen, Rikke Lewinski, Mikko Kuokkanen, Heli Rasinpera, Hatem El-Shanti, Jeong Kee Seo, Michael Alifrangis, Insaf F. Khalil, Abdrazak Natah, Ahmed Ali, Sirajedin Natah, David

Comas, S. Qasim Mehdi, Leif Groop, Else Marie Vestergaard, Faiqa Imtiaz, Mohamed S. Rashed, Brian Meyer, Jesper Troelsen, and Leena Peltonen. "Independent Introduction of Two Lactase Persistence Alleles into Human Populations Reflects Different Histories of Adaptation to Milk Culture." *American Journal of Human Genetics* 82 (2008): 57–72.

Eurobarometer. "Social Values, Science, and Technology," 2005. http://ec.europa.eu/public_opinion/archives/ebs/ebs_225_report_en.pdf (accessed January 3, 2010).

Facebook. http://www.facebook.com/statistics (accessed September 1, 2011). Note: at time of publication, this page was inactive.

Ferguson, Niall. "The Mash of Civilizations." Daily Beast, April 10, 2011. http://www.thedailybeast.com/newsweek/2011/04/10/the-mash-of-civilizations.html (accessed August 25, 2011).

Fineman, Howard. *The Thirteen American Arguments*. New York: Random House, 2008.

Finn, Julian K., Tom Tregenza, and Mark D. Norman. "Defensive Tool Use in a Coconut-Carrying Octopus." *Current Biology* 19 (2009): R1069–70.

Fodor, Jerry, and Massimo Piatelli-Palmarino. *What Darwin Got Wrong*. New York: Farrar, Straus and Giroux, 2010.

Frail, T. A. "Poll: Americans Predict Life in 2050." *Smithsonian Magazine*, August 2010. http://www.smithsonianmag.com/specialsections/40th-anniversary/Poll-Americans-Predict-Life-in-2050.html (accessed June 27, 2011).

Freud, Sigmund. *The Future of an Illusion*. New York: Norton, 1961.

Fukuyama, Francis. Address given to the World Affairs Council, San Francisco, April 20, 2011.

Futuyma, Douglas J. *Evolutionary Biology*. 2nd ed. Sunderland, MA: Sinauer, 2009.

Gabriel, Bonnie. *The Fine Art of Erotic Talk*. New York: Bantam Books, 1998.

Galbraith, John Kenneth. *The New Industrial State*. Harmondsworth, UK: Penguin/Hamish Hamilton, 1967.

Garfinkel, Harold. *Studies in Ethnomethodology*. Englewood Cliffs, NJ: Prentice-Hall, 1967.

Gazzaniga, Michael S. *Nature's Mind*. New York: Basic Books, 1992.

Geertz, Clifford. *The Interpretation of Cultures*. New York: Basic Books, 1973.

Gegax, T. Trent, Catharine Skipp, Jamie Reno, Joan Raymond, and Jerry Adler. "Doubting Darwin." Daily Beast, February 6, 2005. http://www.thedailybeast.com/newsweek/2005/02/06/doubting-darwin.html (accessed February 22, 2005).

Goddard, Matthew R., Charles Godfray, and Austin Burt. "Sex Increases the Efficacy of Natural Selection in Experimental Yeast Populations." *Nature* 434 (March 31, 2005): 636–40.

Goldin, Claudia, and Lawrence F. Katz. *The Race between Education and Technology*. Cambridge, MA: Belknap, 2010.

Gomes, Christina M., and Christopher Boesch. "Wild Chimpanzees Exchange Meat for Sex in a Long-Term Basis." *PLoS ONE* 116 (2009): 1371.

Gould, Stephen Jay. *The Structure of Evolutionary Theory*. Cambridge, MA: Harvard University Press, 2002.

Gould, Stephen Jay, and Richard C. Lewontin. "The Spandrels of San Marco and the Panglossian Paradigm." *Proceedings of the Royal Society of London* 205 (1979): 281–88.

Grayling, A. C. *Against All Gods*. London: Oberon, 2007.

———. "Atheists on Religion." Address given to the London School of Economics, May 12, 2010.

Greene, Brian. *The Hidden Reality*. New York: Alfred A. Knopf, 2011.

Grice, H. Paul. "Logic and Conversation." In *Syntax and Semantics, Vol. 3: Speech Acts*. Edited by P. Cole and J. Morgan, pp. 43–58. New York: Academic Press, 1975.

———. "Meaning." *Philosophical Review* 64 (1957): 377–88.

Griffin, Donald R. *Animal Minds*. Chicago: University of Chicago Press, 2001.

———. *Animal Thinking*. Cambridge, MA: Harvard University Press, 1984.

Grüter, Christoph, M. Sol Balbuena, and Walter M. Farina. "Informational Conflicts Created by the Waggle Dance." *Proceedings of the Royal Society B: Biological Sciences* 275, no. 1640 (June 7, 2008). http://rspb.royalsocietypublishing.org/content/275/1640/1321 (accessed April 1, 2009).

Haldane, J. B. S. "Population Genetics." *New Biology* 18 (1955): 34–51.

Hamilton, William D. "The Genetical Evolution of Social Behavior, Vols. 1 and 2." *Journal of Theoretical Biology* 7 (1964): 1–52.

———. *Narrow Roads of Gene Land: The Collected Papers of W. D. Hamilton, Volume 1: The Evolution of Social Behavior*. Oxford, UK: W. H. Freeman/Spektrum, 1996.

———. *Narrow Roads of Gene Land: The Collected Papers of W. D. Hamilton, Volume 2: The Evolution of Sex*. Oxford: Oxford University Press, 2001.

Harpending, Henry C. *The 10,000 Year Explosion*. New York: Basic Books, 2009.

Harris, Lee. *The Suicide of Reason*. New York: Basic Books, 2007.

Harris, Sam. *The End of Faith*. New York: Norton, 2005.

———. *Letter to a Christian Nation*. New York: Knopf, 2006.

Harris, Sam., Jonas T. Kaplan, Ashley Curiel, Susan Y. Bookheimer, Marc Iacoboni, and Mark S. Cohen. "The Neural Correlates of Religious and Nonreligious Belief." *PLoS ONE*, June 29, 2009. http://www.plosone.org/article/info:doi/10.1371/journal.pone.0007272 (accessed February 9, 2011).

Hart, Donna, and Robert W. Sussman. *Man the Hunted*. Boulder, CO: Westview, 2005.

Hauser, Marc D. *The Evolution of Communication*. Cambridge, MA: MIT Press, 1996.

———. *Moral Minds*. New York: HarperCollins, 2007.

Hawking, Stephen, and Leonard Mlodinow. *The Grand Design*. New York: Bantam Books, 2010.

Hawks, John, Eric T. Wang, Gregory M. Cochran, Henry C. Harpending, and Robert K. Moyzis. "Recent Acceleration of Human Adaptive Evolution." *Proceedings of the National Academy of Sciences* 104 (2007): 20753–58.

Hediger, Heini. "Proper Names in the Animal Kingdom." *Cellular and Molecular Sciences* 32 (1976): 1357–64.

Henshilwood, Christopher S., Francesco d'Errico, Karen L. van Nieker, Yvan Coquinot, Zenobia Jacobs, Stein-Erik Lauritzen, Michel Menu, Renata Garcia-Moreno. "A 100,000-Year Old Ochre-Processing Workshop at Blombos Cave, South Africa. *Science* 14 (October, 2011): 219–22.

Hinde, Robert A. *Religion and Darwinism*. London: British Humanist Association, 1997.

Hirsi Ali, Ayaan. "Book Forum: Infidel." Address presented to the American Enterprise Institute. Washington, DC, February 13, 2007.

———. "From Islam to America: A Personal Journey through the Clash of Civilizations." Address given to the World Affairs Council. San Francisco, CA, May 26, 2010.

———. *Infidel*. New York: Free Press, 2007.

———. *Nomad*. New York: Free Press, 2010.

Hitchens, Christopher. *God Is Not Great*. New York: Twelve, 2007.

"Hitler Jewish? DNA Tests Show Hitler May Have Had Jewish and African Roots." *Huffington Post*, August 8, 2005. http://www.huffingtonpost.com/2010/08/25/hitler-jewish-dna-tests-s_n_693568.html, August 8, 2005 (accessed December 25, 2010).

Hockings, Kimberly J., Tatyana Humle, James R. Anderson, Dora Biro, and Claudia Sousa. "Chimpanzees Share Forbidden Fruit." *PLoS ONE*, September 9, 2007. http://www.plosone.org/article/info%3Adoi%2F10.1371%2Fjournal.pone.0000886 (accessed January 19, 2012).

Hölldobler, Bert, and Edward O. Wilson. *The Superorganism: The Beauty, Elegance, and Strangeness of Insect Societies*. New York: Norton, 2008.

Holmes, Bob. "Did Prehistoric Chimps Use Stone Tools Too?" *New Scientist*, February 12, 2007. http://www.newscientist.com/article/dn11165-did-prehistoric-chimps-use-stone-tools-too.html (accessed April 16, 2007).

Hrdy, Sarah Blaffer. *Mothers and Others: The Evolutionary Origins of Mutual Understanding*. Cambridge, MA: Harvard University Press, 2010.

Hull, David L. *Science as a Process*. Chicago: University of Chicago Press, 1988.

Human Security Centre. *Human Security Report 2005: War and Peace in the 21st Century*. New York: Oxford University Press, 2006.

Hunt, Lynn. *Inventing Human Rights*. New York: Norton, 2007.

Huntington, Samuel P. *The Clash of Civilizations and the Remaking of World Order*. New York: Simon & Schuster, 1996.

Iacoboni, Marco. *Mirroring People*. New York: Farrar, Straus and Giroux, 2008.

Ignatieff, Michael. "Is Nothing Sacred? The Ethics of Television." *Daedalus* 114 (1985): 57–78.

Inglehart, Ron, and Pippa Norris. "The True Clash of Civilizations." *Foreign Policy* (March–April, 2003): 63–70.

"Is It Possible Megan Fox Is Overexposed?" *San Jose Mercury News*, September 2, 2009.

Jackson, P. "Man versus Man-Eater." In *Great Cats: Majestic Creatures of the Wild*. Edited by John Seidensticker, Susan Lumpkin, and Francis Knight, pp. 212–13. Emmaus, PA: Rodale, 1991.

James, William. *Varieties of Religious Experience. A Study of Human Nature*. Harmondsworth, UK: Penguin, 1982.

Jensen, Robert, and Emily Oster. "The Power of TV: Cable Television and Women's Status in India." *Journal of Economics* 124 (2009): 1057–94.

Johanson, Donald, and Blake Edgar. *From Lucy to Language*. New York: Simon & Schuster, 2006.

Johnson, Steven. *Where Good Ideas Come From*. New York: Riverhead Books, 2010.

Jolly, Alison. *Lucy's Legacy: Sex and Evolution in Human Intelligence*. Cambridge, MA: Harvard University Press, 1999.

Jonaitis, Aldona, and Douglas Cole. *Chiefly Feasts: The Enduring Kwakiutl Potlatch*. Seattle: University of Washington Press, 1991.

Jones, Jeffrey M. "Few Americans Oppose National Day of Prayer." Gallup, May 5, 2010. http://www.gallup.com/poll/127721/Few-Americans-Oppose-National-Day-Prayer.aspx (accessed October 2, 2011).

Jones, Steve. *Darwin's Ghost: The Origin of Species Updated*. New York: Random House, 2000.

Judson, Olivia. "An Evolve-By Date." Opinionator, *New York Times*, November 24, 2009. http://opinionator.blogs.nytimes.com/2009/11/24/an-evolve-by-date (accessed September 1, 2011).

Jurmaine, Robert, Lynn Kilgore, Wenda Trevathan, and Russell Ciochon. *Introduction to Physical Anthropology*. Belmont, CA: Wadsworth, 2007.

Kaplan, Hillard, and Kim Hill. "Hunting Ability and Reproductive Success among Male Ache Foragers." *Current Anthropology* 26 (1985): 131–33.

Kauffman, Stuart A. *Investigations*. New York: Oxford University Press, 2000.

Kealey, Terence. "Why Do Men Find Big Lips and Little Noses So Sexy? I'll Paint You a Picture." *Times*, November 28, 2005. http://www.times.co.uk/tto/law/columnists/article/2048633ece (accessed August 12, 2009).

Kelly, Kevin. *What Technology Wants*. New York: Viking, 2010.

Kennedy, Paul. *Preparing for the Twenty-First Century*. New York: Random House, 1993.

Khanna, Parag. "Beyond City Limits." *Foreign Policy* (September 2010): 120–28.

King, Barbara J. *Evolving God*. New York: Doubleday, 2007.

Kitzmiller v. Dover Area School District, United States District Court, M.D., Pennsylvania, December 20, 2005.

Koenig, Walter D., Eric L. Walters, and Joey Haydock. "Variable Helper Effects, Ecological Conditions, and the Evolution of Cooperative Breeding in the Acorn Woodpecker." *American Naturalist* 178 (2011): 145–58.

Koyré, Alexandre. *From the Closed World to the Infinite Universe*. Radford, VA: Wilder, 2009.

Krebs, John. "From Intellectual Plumbing to Arms Race." Presented to the London School of Economics and Political Science, March 16, 2006.

Krukk, Hans. *The Spotted Hyena. A Study of Predation and Social Behavior*. Chicago: University of Chicago Press, 1972.

Kuhn, Steven L., and Mary C. Stiner. "What's a Mother to Do?" *Current Anthropology* 47, no. 6 (2006): 953–80.

Leakey, Louis. *Adam's Ancestors*. London: Methuen, 1934.

LeBlanc, Steven A. *Constant Battles*. New York: St. Martin's, 2003.

Lessig, Lawrence. *Remix: Making Art and Commerce Thrive in the Hybrid Economy*. New York: Penguin, 2008.

Lewis, Bernard. *What Went Wrong?* London: Weidenfeld & Nicolson, 2002.

Lewontin, Richard. "It's Even Less in Your Genes." *New York Review of Books*. May 26, 2001, pp. 23–25.

Lindsay, Steven R. *Handbook of Applied Dog Behavior and Training*. Vol. 1. Ames: Iowa State University Press, 2000.

Lovejoy, Arthur. *The Great Chain of Being*. Cambridge, MA: Harvard University Press, 1936.

Lull, James, *China Turned On: Television, Reform, and Resistance*. London: Routledge, 1991.

———, ed. *Culture in the Communication Age*. London: Routledge, 2001.

———. *Culture-on-Demand*. Oxford, UK: Wiley-Blackwell, 2007.

———. *Media, Communication, Culture*. Cambridge: Polity, 2000.

———, ed. *World Families Watch Television*. Newbury Park, CA: Sage, 1988.

Lull, James, and Stephen Hinerman, eds. *Media Scandals*. Cambridge: Polity, 1997.

Lyell, Charles. *Principles of Geology*. London: John Murray, 1837.

Malinowski, Bronisław. *The Argonauts of the Western Pacific*. London: Routledge & Kegan Paul, 1922.

———. "Culture." In *The Encyclopedia of Social Sciences*. Vol. 6. New York: Macmillan, 1931, pp. 621–46.

Malthus, Thomas R. *An Essay on Population, or A View of Its Past and Present Effects on Human Happiness*. London: John Murray, 1798.

Manji, Irshad. *Allah, Liberty and Love*. New York: Free Press, 2011.

———. *The Trouble with Islam Today*. New York: St. Martin's, 2005.

Marsden, Paul. "Memetics and Social Contagion: Two Sides of the Same Coin?" *Journal of Memetics: Evolutionary Models of Information Transmission* 2 (1998). http://cfpm.org/jom-emit/1998/vol2/marsden_p.html (accessed November 30, 2010).

Marx, Karl, and Friedrich Engels. *The Communist Manifesto*. New York: Monthly Review, 1964.

———. *Manifest der Kommunistischen Partei*. London: Communist League, 1848.

Maynard Smith, John. *Evolution and the Theory of Games*. Cambridge: Cambridge University Press, 1982.

———. *The Theory of Evolution*. Canto ed., rev. Cambridge: Cambridge University Press, 1993.

Maynard Smith, John, and G. R. Price. "The Logic of Animal Conflict." *Nature* 246 (1973): 15–18.

Maynard Smith, John, and Eörs Szathmáry. *The Major Transitions in Evolution*. Oxford, UK: W. H. Freeman/Spektrum, 1995.

———. *The Origins of Life*. Oxford: Oxford University Press, 1999.

Mayr, Ernst. *The Growth of Biological Thought: Diversity, Evolution, and Inheritance*. Cambridge, MA: Belknap, 1982.

McBrearty, Sally, and Alison S. Brooks. "The Revolution That Wasn't: A New Interpretation of the Origin of Modern Human Behavior." *Journal of Human Evolution* 39 (2000): 453–563.

McCracken, Grant. *Culture and Consumption*. Bloomington: Indiana University Press, 1990.

McGrew, William C. "Chimpanzee Technology." *Science* 328 (2010): 579–80.

McLuhan, Marshall. *The Gutenberg Galaxy*. Toronto, ON: Toronto University Press, 1962.

———. *Understanding Media*. New York: New American Library, 1964.

McNally, Terrence. "Atheist Richard Dawkins on 'The God Delusion.'" AlterNet, January 18, 2007. http://www.alternet.org/story/46566/.

McPherron, Shannon P., Zeresenay Alemseged, Curtis W. Marean, Jonathan G. Wynn, Deneé Reed, Denis Geraads, René Bobe, Hamdallah A. Béarat. "Evidence for Stone-Tool-Assisted Consumption of Animal Tissues before 3.39 Million Years Ago at Dikka, Ethiopia." *Nature* 466 (August 2010): 857–60.

Menasce Horowitz, Juliana. "Declining Support for bin Laden and Suicide Bombing." Pew Global Attitudes Project, Pew Research Center Publications, September 10, 2009. http://pewresearch.org/pubs/1338/declining-muslim-support-for-bin-laden-suicide-bombing (accessed August 1, 2011).

Miller, Geoffrey. *Spent: Sex, Evolution, and Consumer Behavior*. New York: Viking, 2009.

Miller, Lisa. "Is God Real?" *Newsweek*, April 9, 2007, pp. 31–37.

Mirsky, Steven. "What's Good for the Group." *Scientific American* (January 2009): 51.

Mithen, Steven. *The Singing Neanderthals*. Cambridge, MA: Harvard University Press, 2005.

Nasr, Vali. *Forces of Fortune: Rise of the New Muslim Middle Class and What It Will Mean for Our World*. New York: Free Press, 2009.

Nass, Clifford, and Li Gong. "Speech Interfaces from an Evolutionary Perspective: Social Psychological Research and Design Implications." *Association for Computing Machinery* 43 (2000): 36–43.

National Center for Social Research. "British Social Attitudes Survey, 25th Report." NatCen, January 2009. http://www.natcen.ac.uk/study/british-social-attitudes-25th-report (accessed January 30, 2007).

Nature. Vol. 455 (October 23, 2008): 1007–1008.

Neff, Bryan D., and Joanna S. Lister. "Genetic Life History Effects on Juvenile Survival in the Bluegill." *Journal of Evolutionary Biology* 20 (2007): 517–25.

Neiva, Eduardo. *Communication Games*. Berlin: Mouton de Gruyter, 2007.

Nettle, Daniel, and Helen Clegg. "Schizotypy, Creativity, and Mating Success in Humans." *Proceedings of the Royal Society B: Biological Sciences* 273 (2006): 611–15.

Newport, Frank. "On Darwin's Birthday, Only 4 in 10 Believe in Evolution." Gallup, February 11, 2009. http://www.gallup.com/poll/114544/Darwin-Birthday-Believe-Evolution.aspx. (accessed February 12, 2009).

———. "Four in 10 Americans Believe in Strict Creationism." Gallup, December 17, 2010. http://www.gallup.com/poll/145286/Four-Americans-Believe-Strict-Creationism.aspx (accessed July 6, 2011).

———. "Near Record High See Religion Losing Its Influence in America." Gallup, December 29, 2010. http://www.gallup.com/poll/145409/Near-Record-High-Religion-Losing-Influence-America.aspx (accessed June 1, 2011).

Nowak, Martin A., Corina E. Tarnita, and Edward O. Wilson. "The Evolution of Eusociality." *Nature* 466 (August 26, 2010): 1057–62.

Nye, Joseph S., Jr. *The Future of Power*. New York: Public Affairs Books, 2011.

Onishi, Norimitsu. "Trying to Save Wild Tigers by Rehabilitating Them." *New York Times*, April 21, 2010. http://www.nytimes.com/2010/04/22/world/asia/22tigers.html (accessed December 23, 2010).

Osvath, Mathias. "Spontaneous Planning for Future Stone Throwing by a Male Chimpanzee." *Current Biology* 19 (2009): R190–91.

Page, Scott. *The Difference: How the Power of Diversity Creates Better Groups, Firms, Schools, and Societies*. Princeton, NJ: Princeton University Press, 2007.

Palin, S., with Lynn Vincent. *Going Rogue: An American Life*. New York: HarperCollins, 2009.

Parr, Lisa A., and Bridget M. Waller. "Understanding Chimpanzee Facial Expression: Insights into the Evolution of Communication." *Social Cognitive and Affective Neuroscience* 1 (2006): 221–28.

Paul, Gregory S. "The Big Religion Questions Finally Solved." *Free Inquiry* (December–January 2008/2009): 24–36.

Pew Forum on Religion & Public Life. "Faith in Flux: Changes in Religious Affiliation in the U.S.," April 27, 2009. http://www.pewforum.org/Faith-in-Flux-Changes-in-Religious-Affiliation-in-the-US.aspx (accessed October 30, 2009).

———. "Many Americans Mix Multiple Faiths," December 9, 2009. http://www.pewforum.org/Other-Beliefs-and-Practices/Many-Americans-Mix-Multiple-Faiths.aspx (accessed January 30, 2012).

———. "The Religious Composition of the 112th Congress." Faith on the Hill, February 28, 2011. http://www.pewforum.org/Government/Faith-on-the-Hill-The-Religious-Composition-of-the-112th-Congress.aspx (accessed November 22, 2011).

———."U.S. Religious Landscape Survey," June 23, 2008. http://religions.pewforum.org/reports (accessed June 23, 2008).

Pew Global Attitudes Project. "Global Digital Communication: Texting Social Networking Popular Worldwide." Pew Research Center, December 20, 2011. http://www.pewglobal.org/2011/12/20/global-digital-communication-texting-social-networking-popular-worldwide/ (accessed January 8, 2012).

———. "The Global Middle Class." Pew Research Center, February 12, 2009. http://www.pewglobal.org/2009/02/12/the-global-middle-class/ (accessed December 23, 2010).

———. "Global Publics Embrace Social Networking." Pew Research Center, December 15, 2010. http://www.pewglobal.org/2010/12/15/global-publics-embrace-social-networking (accessed March 25, 2011).

———. "Mixed Views of Hamas and Hezbollah in Largely Muslim Nations." Pew Research Center, February 4, 2010. http://www.pewglobal.org/2010/02/04/mixed-views-of-hamas-and-hezbollah-in-largely-muslim-nations (accessed February 3, 2012).

———. "Muslim Publics Divided on Hamas and Hezbollah." Pew Research Center, December 2, 2010. http://www.pewglobal.org/2010/12/02/muslims-around-the-world-divided-on-hamas-and-hezbollah/ (accessed October 18, 2011).

———. "Widespread Support for Banning Full Islamic Veil in Western Europe." Pew Research Center, July 8, 2010. http://www.pewglobal.org/2010/07/08/widespread-support-for-banning-full-islamic-veil-in-western-europe/ (accessed November 8, 2010).

———. "World Publics Welcome Global Trade—But Not Immigration." Pew Research Center, October 4, 2007. http://www.pewglobal.org/2007/10/04/world-publics-welcome-global-trade-but-not-immigration/ (accessed April 18, 2009).

Pew Internet and American Life Project. "On MySpace, Girls Seek Friends, Boys Flirt," January 7, 2007. http://pewinternet.org/Media-Mentions/2007/On-MySpace-girls-seek-friends-boys-flirt-study.aspx (accessed February 23, 2012).

Phelps, Andrea C., Paul K. Maciejewski, Matthew Nilsson, Tracy A. Balboni, Alexi A. Wright, M. Elizabeth Paulk, Elizabeth Trice, Deborah Schrag, John R. Peteet, Susan D. Block, and Holly G. Prigerson. "Use of Religious Faith to Cope with Advanced Cancer Associated with Receiving Medical Care Near Death." *Journal of the American Medical Association* 301 (2009): 1140–47.

Pinker, Steven. *The Better Angels of Our Nature: Why Violence Has Declined.* New York: Viking, 2011.

———. "A History of (Non) Violence." *Foreign Policy* (December 2011).

———. *The Language Instinct.* New York: William Morrow, 1994.

———. "The Moral Instinct." *New York Times Magazine*, January 13, 2008, pp. 32–37, 52, 55–58.

Pinker, Steven, and Paul Bloom. "Natural Language and Natural Selection." *Behavioral and Brain Science* 13 (1990): 707–84.

Pontzer, Herman, David A. Raichlen, and Michael D. Sockol. "The Metabolic Cost of Walking in Humans, Chimpanzees, and Early Hominins." *Journal of Human Evolution* 56 (2009) 43–54.

Post, Stephen G., ed. *Altruism and Health.* Oxford: Oxford University Press, 2007.

Price, G. R. "Selection and Covariance." *Nature* 227 (1970): 520–21.

"Public Acceptance of Evolution in Science." National Center for Science Education August 15, 2006. http://ncse.com/news/2006/08/public-acceptance-evolution-science-00991 (accessed January 30, 2007).

Public Broadcasting Service. "Nature: What Females Want and Males Will Do," April 6 and 13, 2008.

Putnam, Robert D. "*E Pluribus Unum*: Diversity and Community in the 21st Century." *Scandinavian Political Studies* 30 (2007): 137–74.

Putnam, Robert D., and Donald E. Campbell. *American Grace: How Religion Divides and Unites Us*. New York: Simon & Schuster, 2010.

Quammen, David. *The Reluctant Mr. Darwin*. New York: Norton, 2006.

Richerson, Peter J., and Robert Boyd. *Not by Genes Alone*. Chicago: Chicago University Press, 2005.

Rifkin, Jeremy. *The Empathic Civilization*. New York: Tarcher, 2009.

Rubenstein, Dustin R., and Irby J. Lovette. "Temporal Environmental Variability Drives the Evolution of Cooperative Breeding in Birds." *Current Biology* 17 (2007): 1414–19.

Rushdie, Salman. Comment on "Bill Moyers on Faith and Reason." Public Broadcasting Service, June 23, 2006.

Russell, Bertrand. "Is There a God?" Campaign for Philosophical Freedom. http://www.cfpf.org.uk/articles/religion/br/br_god.html.

———. *Why I Am Not a Christian*. New York: Simon & Schuster, 1957.

Ryan, Christopher, and Cacilda Jethá. *Sex at Dawn: The Prehistoric Origins of Modern Sexuality*. New York: Harper, 2010.

Schelling, Thomas C. *The Strategy of Conflict*. Cambridge, MA: Harvard University Press, 1960.

Schopp, Ludwig. *The Fathers of the Church*. Vol. 2. Internet Archive. http://www.archive.org/details/fathersofthechur009935mbp (accessed September 4, 2011).

Schumpeter, Joseph A. *Capitalism, Socialism, and Democracy*. New York: Routledge, 2006.

Schusterman, Ronald J., Colleen R. Kastak, and David Kastak. "The Cognitive Sea Lion: Meaning and Memory in the Laboratory and Nature." In *The Cognitive Animal: Empirical and Theoretical Perspectives in Animal Cognition*. Edited by Marc Berkoff, Collin Allen, and Gordon M. Burghardt, pp. 256–57. Cambridge, MA: MIT Press, 2002.

Schwab, George. *The Challenge of the Exception*. Berlin: Duncker & Humblot, 1989.

Seabright, Paul. *The Company of Strangers*. Princeton, NJ: Princeton University Press, 2004.

Sebeok, Thomas A. "Naming in Animals, with Reference to Playing." *Semiotic Inquiry* 1 (1981): 121–35.

Seyfarth, R. M., and D. L. Cheney. "Signalers and Receivers in Animal Communication." *Annual Review of Psychology* 54 (2003): 145–73.

Shapiro, Michael. "Who Killed Pat Tillman?" AlterNet, June 13, 2007. http://www.alternet.org/world/53827 (accessed January 3, 2009).

Singer, Peter. *A Darwinian Left: Politics, Evolution, and Cooperation*. (New Haven, CT: Yale University Press, 2000.

Sloterdijk, Peter. *Le Palais de Cristal*. Paris: Maren Sell Éditeurs, 2006.

Sober, Elliott, and David Sloan Wilson. *Unto Others: Evolution and the Psychology of Selfish Behavior*. Cambridge, MA: Harvard University Press, 1998.

Sowell, Thomas. *Race and Culture*. New York: Basic Books, 1994.

Smith, Adam. *The Wealth of Nations*. New York: Bantam, 2003.

Smith, Eric Alden. "Why Do Good Hunters Have Higher Reproductive Success?" *Human Nature* 15 (2004): 343–64.

Smith, Zadie. "The Social Network." *New York Review of Books* 25 (2010): 57–60.

Stanford, Craig B. *The Hunting Apes: Meat Eating and the Origins of Human Behavior*. Princeton, NJ: Princeton University Press, 1999.

Stanford Encyclopedia of Philosophy. "Pascal's Wager." http://plato.stanford.edu/entries/pascal-wager/ (accessed January 2, 2010).

Stegner, Wallace. *This Is Dinosaur*. Boulder, CO: Robert Rinehart, 1985.

Stewart, Edward C. "Culture of the Mind." In *Culture in the Communication Age*. Edited by James Lull, pp. 9–30. London: Routledge, 2001.

Stiglitz, Joseph. *Freefall*. New York: Norton, 2010.

Stinchcombe, Arthur J. "Social Structure and Organizations." In *Handbook of Organizations*. Edited by J. G. Marsh, pp. 153–93. New York: Rand McNally, 1965.

Stoddard, Ed. "Poll Finds More Americans Believe in Devil than Darwin." Reuters UK, November 29, 2007. http://uk.reuters.com (accessed August 21, 2008).

Stringer, Chris, and Peter Andrews. *The Complete World of Human Evolution*. New York: Thames & Hudson, 2005.

Takahashi, Toshie. *Audience Studies: A Japanese Perspective*. New York: Routledge, 2010.

Thompson, W. L. "Agonistic Behavior in the House Finch, Part 1: Annual Cycle and Display Patterns." *Condor* 62 (1960).

Tibi, Bassam. *The Challenge of Fundamentalism*. Berkeley: University of California Press, 2002.

Tierney, John. "A World of Eloquence in an Upturned Palm." *New York Times*, August 28, 2007.

Tomasello, Michael. *Origins of Human Communication*. Cambridge, MA: MIT Press, 2008.

———. "What Kind of Evidence Could Refute the UG Hypothesis?" *Studies in Language* 28 (2004): 642–44.

———. *Why We Cooperate*. Boston: MIT Press, 2009.

Trivers, Robert L. "The Evolution of Reciprocal Altruism." *Quarterly Journal of Biology* 46 (1971): 33–57.

———. *The Folly of Fools: The Logic of Deceit and Self-Deception in Human Life.* New York: Basic Books, 2012.

Turkle, Sherry. *Alone Together*. New York: Basic Books, 2011.

Turner, Jonathan H. *On the Origins of Human Emotions*. Stanford, CA: Stanford University Press, 2000.

Tylor, Edward B. *Primitive Culture: Researches into the Development of Mythology, Philosophy, Religion, Art, and Custom*. London: John Murray, 1871.

United Nations Development Programme. *Human Development Report: Cultural Liberty in Today's Diverse World*. New York: Oxford University Press, 2004.

———. *Human Development Report: International Cooperation at a Crossroads.* New York: Oxford University Press, 2005.

Vanhaeren, Marian, Francesco d'Errico, Chris Stringer, Sarah L. James, Jonathan A. Todd, and Henk K. Mienis. "Middle Paleolithic Shell Beads in Israel and Algeria." *Science* 312 (June 23, 2006): 1785–88.

Vieira, Antonio. *Cartas*. Rio de Janeiro: W. W. Jackson, 1948.

"Viral Video: A YouTube Is Born for the Arab World." *Newsweek*, January 22, 2007, p. 8.

von Frisch, Karl. *The Dance Language and Orientation of Bees*. Cambridge, MA: Belknap, 1967.

———. "Honeybees: Do They Use Direction and Distance Information Provided by Their Dancers?" *Science* 158 (1967): 1073–76.

Wade, Nicholas. *Before the Dawn*. New York: Penguin, 2006.

———. "Equality between the Sexes: Neanderthal Women Joined Men in the Hunt." *New York Times*, December 5, 2006.

———. *The Faith Instinct*. New York: Penguin, 2009.

———. "Scientist Finds the Beginnings of Morality in Primate Behavior." *New York Times*, March 20, 2007.

———. "A Speech Gene Reveals Its Bossy Nature." *New York Times*, November 12, 2009.

———. "Still Evolving, Human Genes Tell New Story." *New York Times*, March 7, 2006. http://www.nytimes.com/2006/03/07/science/07evolve.html?scp=1&sq=Still%20Evolving,%20Human%20Genes%20Tell%20New%20Story&st=cse (accessed March 25, 2011).

Wallerstein, Immanuel. *The Decline of American Power*. New York: New Press, 2003.

Washburn, Sherwood L. *The Social Life of Early Man*. London: Methuen, 1962.

Washburn, Sherwood L., and V. Avis. "Evolution of Human Behavior." In *Behavior and Evolution*. Edited by Anne Roe and George Simpson, pp. 421–36. New Haven, CT: Yale University Press, 1958.

Wells, Spencer. *Pandora's Seed*. New York: Random House, 2010.

Welsh Jennifer. "8-Legged Sex Trick? Spiders Give Worthless Gifts, Play Dead," November 14, 2011. http://www.livescience.com/17010-spider-gifts-play-dead-mating.html (accessed February 5, 2012).

Wickler, Wolfgang. *Mimicry in Plants and Animals*. New York: McGraw-Hill, 1968.

Wilkinson, Gerald S. "Reciprocal Food Sharing in the Vampire Bat." *Nature* 389 (1984): 181–84.

Wilkinson, Richard, and Kate Pickett. *The Spirit Level: Why Greater Equality Makes Societies Stronger*. New York: Bloomsbury, 2009.

Williams, George C. *Adaptation and Natural Selection: A Critique of Some Current Evolutionary Thought*. Princeton, NJ: Princeton University Press, 1996.

———. "Measurements of Consociation among Fish and Comments on the Evolution of Schooling." *Publications of the Museum*, Michigan State University, Biological Series 2/7 (1966): 149–84.

Williams, Raymond. *The Long Revolution*. New York: Columbia University Press, 1962.

Wilson, David Sloan. *Darwin's Cathedral*. Chicago: University of Chicago Press, 2002.

———. *Evolution for Everyone*. New York: Delacorte, 2007.

Wilson, David Sloan, and Edward O. Wilson. "Rethinking the Theoretical Foundations of Sociobiology." *Quarterly Review of Biology* 82 (2007): 327–48.

Wilson, Edward O., ed. *From So Simple a Beginning: The Four Great Books of Charles Darwin*. New York: Norton, 2006.

———. *The Insect Societies*. Cambridge, MA: Harvard University Press, 1971.

———. *Sociobiology: A New Synthesis*. Cambridge, MA: Harvard University Press, 1975.

Wootton, David. *Galileo: Watcher of the Skies*. Princeton, NJ: Princeton University Press, 2010.

Worth, Robert F. "Arab TV Tests Societies' Limits with Depictions of Wine, Sex, and Equality." *New York Times*, September 27, 2008.

Wright, Robert. *The Evolution of God*. New York: Little, Brown, 2009.

Wright, Sewall. "The Evolution of Dominance." *American Naturalist* 63 (1929): 556–61.

Wynne-Edwards, V. C. *Animal Dispersion in Relation to Social Behavior*. Edinburgh, Scotland: Oliver and Boyd, 1962.

Yi, Xin. "Sequencing of 50 Human Exomes Reveals Adaptation to High Altitude." *Science* 329 (2010): 75–78.

Yoon, Carol Kaesuk. "Loyal to Its Roots." *New York Times*, June 10, 2008.

Zachary, G. Pascal. *The Global Me*. New York: Public Affairs, 2000.

Zahavi, Amotz, and Avishag Zahavi. *The Handicap Principle: A Missing Piece in Darwin's Puzzle*. Oxford: Oxford University Press, 1997.

Zakaria, Fareed. *The Future of Freedom*. New York: Norton, 2003.

Zimmer, Carl. "Crunching the Data for the Tree of Life." *New York Times*, February 10, 2009.

———. "In Games, an Insight into the Rules of Evolution." *New York Times*, July 31, 2007.

INDEX

Definitions are indicated by **bold** page numbers.